T0130988

Delivering Utility Computing

Delivering Utility Computing

Business-driven IT Optimization

Guy Bunker and Darren Thomson

John Wiley & Sons, Ltd

Published by John Wiley & Sons Ltd, The Atrium, Southern Gate, Chichester, West Sussex
PO19 8SQ, England

Telephone (+44) 1243 779777

Email (for orders and customer service enquiries): cs-books@wiley.co.uk
Visit our Home Page on www.wiley.com

This publication is designed to provide accurate and authoritative information in regard to
the subject matter covered. It is sold on the understanding that the Publisher is not engaged
in rendering professional services. If professional advice or other expert assistance is
required, the services of a competent professional should be sought.

Other Wiley Editorial Offices

John Wiley & Sons Inc., 111 River Street, Hoboken, NJ 07030, USA

Jossey-Bass, 989 Market Street, San Francisco, CA 94103-1741, USA

Wiley-VCH Verlag GmbH, Boschstr. 12, D-69469 Weinheim, Germany

John Wiley & Sons Australia Ltd, 42 McDougall Street, Milton, Queensland 4064, Australia

John Wiley & Sons (Asia) Pte Ltd, 2 Clementi Loop #02-01, Jin Xing Distripark,
Singapore 129809

John Wiley & Sons Canada Ltd, 22 Worcester Road, Etobicoke, Ontario, Canada M9W 1L1

British Library Cataloguing in Publication Data

A catalogue record for this book is available from the British Library

ISBN 0-470-01576-4

Typeset in 11/13pt Palatino by TechBooks, Delhi, India
Printed and bound in Great Britain by Antony Rowe, Chippenham, Wiltshire
This book is printed on acid-free paper responsibly manufactured from sustainable
forestry in which at least two trees are planted for each one used for paper production.

Contents

Time and IT March On

We live in an era of change; not only is change happening all the time, but the rate of change is accelerating and customers' requirements and the environments that provide them are not the only factors affected. While we were writing this book there were a number of large IT industry mergers, not least when the company we both work for, VERITAS Software, merged with Symantec Corporation in July 2005.

Utility Computing is all about delivering IT as a flexible, cost-effective service. One of the key premises is that the service needs to be delivered in a timely and secure manner and when developing a utility infrastructure time needs to be spent putting in place practices and procedures as well as technology to mitigate risk. As a company the new Symantec remains committed to Utility Computing and is now broadening the picture to include all aspects of security, from Intrusion Detection through anti-virus and anti-spam to data encryption. This combination of technology is resulting in a new sector, Information Integrity. The inclusion of security at all points in the IT infrastructure will not be new to administrators around the globe. However, the strong integration of security technologies with availability technologies will give rise to a new generation of products that will enable Utility environments to be rolled out more easily and with greater security than before.

Guy Bunker
Darren Thompson
December 2005

Delivering Utility Computing. Guy Bunker and Darren Thomson
© 2006 VERITAS Software Corporation. All rights reserved.

About the Authors

Dr. Guy Bunker is CTO for the Application and Service Management Division and Distinguished Engineer at VERITAS Software Corporation. He is responsible for the technical vision for utility computing at VERITAS and for running a number of related research projects.

Guy has worked for VERITAS for nearly a decade in a number of different product divisions, most recently leading research into service level management and the use of new technologies in a utility computing environment. He has been a member of a number of industry bodies driving standards in computer storage and management and is currently a member of the Global Grid Forum and the Grid Market Awareness Council.

Guy is a regular presenter at many conferences, including JavaONE, Tivoli World, Linux on Wall Street and the VERITAS user conference, VISION. Guy was also a co-author of the first VERITAS utility computing book: *From Cost Center to Value Center: Making the move to utility computing*.

Prior to VERITAS, Guy worked for a number of companies, including Oracle, where he was the architect for their Business Process Re-engineering tools.

Guy holds a PhD in Artificial Neural Networks from King's College London and is a Chartered Engineer with the IEE.

Darren Thomson works as Worldwide Practice Director, Utility Computing within VERITAS Global Services. He is responsible for the service development and strategic consulting delivery.

Since joining VERITAS in July 2003, Darren has been working closely with many global customers to help them to realize their utility computing visions. Before joining VERITAS, Darren worked at The Morse Group, a European systems integrator focused on the design and implementation of critical IT systems in the Financial Services, Telco and Media industries. His final position at Morse was as their Group Technical Strategist, focused on 'next generation' server and storage technologies. This role brought Darren into contact with many of today's leading edge companies such as Egenera, VMWare, EMC, Platform Computing and Datacore. A citizen of the UK, Darren was educated in Hertfordshire, England and now holds several IT related certifications, such as the Total Cost of Ownership Expert qualification from The Gartner Group.

Foreword[1]

In 2001 the technology sector took a major hit in the financial mar-
kets. I remember the year because my modest investment portfolio
lost half of its value. The decline was first felt by dot.com companies
that went bust as the market showed no mercy on firms with small to
non-existent revenue streams, but the effects quickly spread across
the entire industry. Internal Information Technology (IT) organizations began to feel the crunch as the budget axe fell upon projects
and ongoing IT operations. At the International Monetary Fund
(IMF or "Fund") where I lead the Server & Storage Infrastructure
Team, the IT department was ill-prepared to defend its budgets as
Fund management demanded increasing fiscal accountability. Each
of you that works in Technology has surely felt the ramifications of
the technology sector's fall. Our challenge is to restore faith in our
industry and to reposition IT from a cost center to a value center.

Stagnant or decreasing IT budgets have not necessarily resulted in
flat or decreasing demands on IT departments. While the "do more
with less" mantra has become cliché to the point of annoyance, it is
nonetheless reality for most IT professionals. By way of example, the
Server & Storage Infrastructure Team at the IMF is responsible for
maintaining twice as many servers and four times as much electronic
storage in 2005 as it did in 2001, with the same level of funding.

---

[1] The views expressed in this foreword are those of the author and should not be attributed
to the International Monetary Fund, its Executive Board, or its management.

Meanwhile, user tolerance for outages has evaporated and security patching of our 400 Windows servers has become a monthly ritual.

To "do more with less" requires efficiency gains, which may seem frustratingly beyond one's grasp. At the IMF, my colleagues and I struggled to find a solution to our multifaceted dilemma, which is characterized by the items in the list below. We learned through external consultants and dicussions with peers in other organizations that our predicament is not unique, and that most medium to large IT organizations suffer from similar problems.

- Far too many people from across the IT organization have full administrative privileges in our Windows server environment;
- Few people outside of the IT Infrastructure group understand the services we provide;
- There is an inability to correlate our service offerings to a meaningful cost structure;
- Poor asset utilization exists – e.g. CPU and disk utilization are, on average, very low;
- IT project teams are able to acquire a low-end server for $8,000, with no responsibility for the annual server TCO of $15,000 (i.e. no exposure of ongoing operational costs);
- In the absence of Service Level Agreements (SLAs), project managers are hesitant to cede control of any part of the application stack (from the hardware to application layer);
- Human error contributes to 80-90% of service outages;
- In the absence of a structure that maps quality-of-service to cost, customers always choose "First-Class" service.

To address these, and similar problems across the IT organization, a plethora of initiatives have been undertaken. ITIL-based service management for change, release, incident, and problem management; a High Availability (HA) Program, and a server consolidation project represent some of main the initiatives affecting the team. Yet a more comprehensive approach is needed to tie these efforts together, optimize the value of these efforts, and ensure that ALL of our major problems are addressed. The cornerstone that binds our efforts together is a utility computing initiative.

Dr. Guy Bunker and Darren Thomson, utility computing thought leaders, have developed a methodology that is predicated on a perfect blend of concept and real-world experience. This book will give you the conceptual knowledge and practical guidance to understand and implement utility computing in your organization whether you work in an internal, or a service provider, IT organization. Whether you are a CFO looking at the bottom line for your business, a CIO aligning IT with business goals, a system architect designing IT services, or an IT manager like me trying to move beyond survival mode to proactive management of your IT environment, this book offers a practical guide for transforming IT to a service-led organization.

At the Fund, the engineers in the Server & Storage Infrastructure Team have used the utility computing principles that Dr. Bunker and Mr. Thomson describe in detail in this book. Our IT Utility Service Environment (IT USE) initiative is quickly gaining traction because of the concrete objectives we were able to define using the model developed by the authors. We expect to realize clearly demonstrated cost savings through IT USE within two years. This compares favorably with service management initiatives (e.g. ITIL), which have been underway for several years, and will take several more to fully implement at the Fund. IT USE ensures that well-defined processes around our HA program are established, and the necessary controls over the server and storage environment to achieve HA, are implemented. We are confident that the shared IT USE infrastructure will help us to realize significantly higher resource utilization. More importantly, by addressing the process by which servers and storage are procured, deployed, maintained and, most importantly, costed, ensures that efficiency gains will be sustained.

If you are interested in IT organizational transformation that properly blends people and process while leveraging advances in technology (e.g. virtualization, workflow automation, provisioning, etc.) this book is definitely for you. The ultimate benefit to be derived from this book, however, is the repositioning of your IT organization as a value center and the establishment of a complementary relationship with the business side of your organization.

Thomas J. Ferris
Senior Information Technology Officer
International Monetary Fund

Acknowledgements

This book could not have been written without the wholehearted support of our colleagues at work, or our managers, who turned a blind eye to the odd hour or two spent writing when we should have been doing something else. The list of colleagues is so long that we are bound to have forgotten someone, in which case our apologies. Here, for the record, are the ones who have proven to be most memorable: Bob Adair, Mark Bregman, Christopher Chandler, Cary Christopherson, Bill Forsyth, Peter Jeffe, Abhijit Kale and the team in Pune, Tom Lanzatella, Paul Massiglia, Chuck Palczak, Susan Rutherford, Bob Santiago, Kaajal Shaikh, Mike Spink, Mike Tardif and Charlie Van Meter. Finally, thanks to the many customers we have spoken to, who have helped us refine the methodology, and to Rob Craig for the proof review – the comments you made helped us to remove many of the sharp edges.

Thanks also go to Birgit, Joanna and Julie at Wiley for being so patient with us as we learned the ropes of publishing.

Darren would like to extend a thank-you to family and friends for tolerating the various mood swings and seemingly endless conversations about the trials of writing your first book. Thanks also to the many colleagues and acquaintances that 'kept me on track' and encouraged me through this experience. In particular to the utility computing practitioners who work as part of my consulting practice. A great deal of credit must be extended, in particular, to Mike Spink, who designed and authored much of the transformational method shown in this book. Thanks, Mike . . . you are an inspiration! Lastly, thanks to Guy. It would have been a much bumpier journey without your experience and guidance.

Guy would like to thank Susie and Veryan for putting up with him while writing this. Writer's block is not just something that the blockbuster author seems to get, and breaking out of it can be 'interesting' for all those around. Thanks also to my team and to Richard Barker, who once said that writing was good for the soul, I'm not sure that that is always the case, but it's certainly quite a ride.

Guy Bunker
Darren Thomson
December 2005

Who Should Read This Book and Why?

This book is for any person who is contemplating transforming the data center into a utility or delivering IT as a service. Covering topics of interest to the CIO through to the IT administrator, its premise is that there is a better way for a business to run IT, and treating it as a customer-facing utility maximizes effectiveness. Simply stated, 'by adopting a utility infrastructure, both cost savings and increased utilization and efficiency of IT can be achieved over traditional methods'.

This book is divided into three parts. The first introduces the topic and discusses why it should happen now. It continues by creating an analogy between running an IT organization (ITO) that delivers IT as a service and how a restaurant is run. The second part describes a complete transformational model and methodology for achieving it, covering all aspects of turning a traditional ITO into a service delivery organization. The final part is dedicated to addressing some of the barriers, cultural implications and adoption strategies for utility computing, and finishes with a look into the future. Appendices are provided, giving case study examples, as well as some helpful template documents for use in transforming the ITO.

List of Figures

Delivering Utility Computing.   Guy Bunker and Darren Thomson
© 2006 VERITAS Software Corporation. All rights reserved.

List of Tables

Part One

Introducing Utility Computing

'Just get a bigger pipe; it can all be solved with a bigger pipe.'
Susanna Sherrell

The premise of this book is that there is now a seed change in the way in which enterprise information services need to be accounted for. Strict alignment with the business is required to ensure that all resources create a suitable return on investment. By looking at other utilities, it is possible to discover the facets that make them successful, and in turn apply these to IT. In this part, we examine how we have arrived at this juncture and look at the way a restaurant is run to find an analogy for a service-driven organization that may be applicable to the next generation of IT organization.

1

Introduction

Upon reflection we will see that the mid-2000s was an inflection point when it came to computing. The paradigm shift of utility computing has come about as information technology (IT) has moved from being *just a business enabler* to becoming *a true business differentiator*. The cost associated with IT has always been seen as high, but necessary and the .com crash in 2000 started putting the squeeze on IT budgets around the world. Alignment with the business has become the new IT mantra and it is up to the CIO to show value rather than just cost (See Figure 1.1).

Pushed by the lines of business to support more applications and their data with greater availability, while the CEO and CFO are squeezing costs, the CIO now has a challenge on his hands. Technology in itself has become cheaper and flexibility in the ways it can be used has increased, these point to improvements for the business. However, the sad fact is that this increasingly complex environment has become very expensive to manage. It is now a well-accepted fact that the business will spend $3 on storage management for every $1 it spends on storage hardware (Nicolett and Paquet, 2001).[1] The heterogeneous environment is here to stay and is now applied not only to proprietary hardware and its associated operating system (OS), but also to storage, commodity hardware

---

[1] While there are no published figures for other aspects of IT, for example server management, the increase in provisioning and reprovisioning of servers and patch management, the numbers would probably be similar. Other, unsubstantiated, numbers put the overall cost of IT management at between $7–$10 for each $1 spent on hardware.

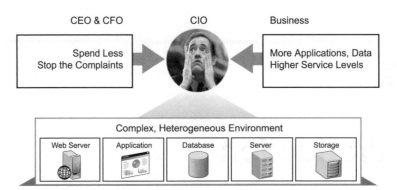

Figure 1.1 The CIO's dilemma.

with interchangeable operating systems and, finally, even to applications, be they on the same OS or in some cases on different ones.

The storage area network (SAN) was held up as the IT environment of the future and in some ways has held up to its promise, but the unforeseen management costs have held it back from universal adoption. Other new technologies, such as iSCSI and Infiniband, which show promise for adding value to the business, are not getting any traction as the cost to manage them is too high and the IT organization (ITO) is already under too much pressure just maintaining what it has already got.

In order to tackle rising costs, the CIO has traditionally turned to outsourcing (Figure 1.2). A large outsource company would come into the business and take the whole of IT off its hands, buying back the equipment, leasing the premises and employing the staff. It

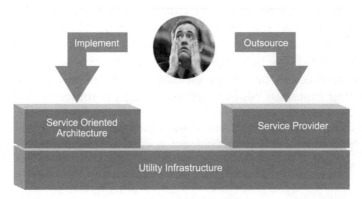

Figure 1.2 The CIO's options.

would then deliver IT as a service for a known and agreed upon price and terms. While this would not necessarily reduce the costs to the absolute minimum, it would, at least in theory, stop the escalation of them, leaving the business to concentrate on its core business and ultimately on making money.

However, utility computing offers the CIO an alternative. By moving to a service oriented architecture (Figure 1.2), the CIO can reduce costs by understanding where the costs actually are and then putting in place an agile ITO that can focus on delivering value to the organization as a whole. This will respond quickly to the changing needs and, through the introduction of automation of best practice, will allow the ITO to concentrate on innovation, thereby helping the business differentiate itself from its competition.

This book introduces utility computing as a better way for a business to run its IT services and puts forwards a complete methodology for transforming the ITO. Fortunately, the move to delivering IT as a service using a utility infrastructure is not an all-or-nothing project, there are a number of stages a business can go through, depending on which services are right for them to start with.

The chapters in Part One cover a general introduction to utility computing, the history of how it came about and the utility model.

Part Two elaborates on the technology and concepts behind utility computing, along with the complete transformational methodology. Step-by-step transition plans are put forward for all the major IT services, together will a method for determining which is best suited to your business.

Finally, Part Three covers the broader implications to the business of a utility computing strategy, how to overcome the cultural implications and develop a successful adoption strategy. It finishes with a look to the future in terms of what might come next.

REFERENCE

Nicolett, M. and Paquet, R. (2001) *The Cost of Storage Management: A Sanity Check*. Gartner Group.

2

What is Utility Computing?

2.1 OVERVIEW

In essence, the concept behind utility computing is simple, an almost utopian dream for the IT organization (ITO)–costs go down while efficiency and effectiveness go up (Figure 2.1). It is a simple message in theory, but in practice, there is the 'so prove it' component, which makes it difficult.

This chapter looks at what utility computing is and how an ITO needs to change in order to support it.

2.2 THE CHANGING ROLE OF IT

Before we begin to understand what utility computing is, we need to examine how the role of IT, and those involved, is changing. Recognizing and acknowledging that there is a change occurring is one componant of creating a successful utility computing strategy. While there are many different people and roles that are linked intrinsically to the success of the IT organization, there are three primary ones:

- the CIO;
- IT consumers, for example the application owner or end user;
- the IT organization itself.

Delivering Utility Computing.   Guy Bunker and Darren Thomson
© 2006 VERITAS Software Corporation. All rights reserved.

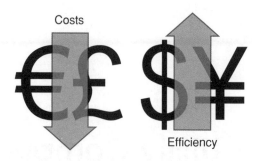

Costs

Efficiency

Figure 2.1 Utility computing, what is it all about?

Satisfying these people, proving to them that the utility approach is better than the existing way of delivering IT, will make the difference between success and failure. As we saw in the previous chapter, the CIO's interest in IT is that, ultimately, the responsibility stops with them. They need to ensure that the business is getting value from IT and that IT is not a barrier to the business. Often, they need high-level information to feed to the CEO and the board, who do not want to hear about numbers of servers or quantities of storage, but want to hear about alignment to the business, business benefits and how IT is being used to accelerate business differentiation.

The consumer of IT, the line of business, the department or the end user, is not interested in IT at all . . . until it breaks. When that happens, they are first on the telephone to find out what is happening and when it will be fixed. The IT organization spends much of its time reacting to problems and issues, rather than planning proactively. Highly skilled system administrators spend their time chasing a multitude of problems which might not necessarily be theirs, but which manifest themselves to the users as a system not being available. The cost-conscious CIO is also looking to charge the consumers of IT for the amount they consume. This dawning era means that the consumer is now interested in what it is they are using, how they are using it, and, most importantly, whether they can stop using it or can get it from someone else more cheaply or faster than using internal resources. While chargeback of IT is not seen by all as a viable business model for running IT, the visibility into usage and the costs associated are. In the past, simple things like backup were taken for granted, no-one worried over how many tapes were in use or how many copies of data they held as the cost associated was not passed on. Like the CIO, the consumer is not interested in every single small detail, but they do want figures to which they can relate.

Finally, the IT organization (ITO) has pressure from above to spend less and stop the complaints, and pressure from below to deliver more applications with improved availability and performance. The business benefit that IT provides is not in dispute, but the lifecycle of new applications is becoming compressed, with applications in some industries only lasting 9–12 months before they either become obsolete or no longer provide the competitive advantage the business requires. In order for a business to make best use of its ideas, the IT organization needs to be able to respond quickly and efficiently to requests from the individual lines of business. In the ideal world, most lines of business would like to be able to do as much as possible themselves rather than having to rely on other departments, self-service portals have become the norm for many everyday tasks, why not for consuming IT? However, due to the complexity of the environment, it is reactive, rather than proactive, work that ends up taking the majority of the ITO's day. Another side effect this pressure causes is that any innovation within the ITO is squeezed out. There is no time to look at new technology or plan for the future. For a business to remain competitive, the innovation needs to return and the role of IT needs to change.

Back to utility computing. The concept is simple, by reorganizing IT and applying some new principles, IT utilization will increase, efficiency will improve and costs will go down. Unfortunately, like the blind men who come across an elephant (Saxe 1964) (see Figure 2.2), everybody has heard the story slightly differently, and because

Figure 2.2   The blind men and the elephant.

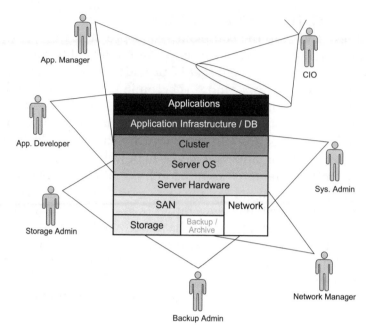

Figure 2.3 IT seen from different perspectives.

everyone is looking at it from a different perspective everyone concludes it to be something different.

Recognizing that utility computing means different things to different people, and understanding that each has a different set of requirements that need to be satisfied from the outset actually makes it easier than trying to put them all together. It is all a question of context and scope (see Figure 2.3). The Application Manager is only interested in his/her application, how it is performing, whether it is being backed up, how much storage it is using and whether the users can see and use the application in a timely manner. The CIO needs to know what the overall IT performance is and, while they may be interested in a particular application, it is the environment as a whole on which they need data. What is the utilization? Is there enough capacity based on current usage trend? What is the overall cost, etc?

2.3 UTILITY COMPUTING

Utility computing is as much about process and principle as it is about technology. The end-game is to put in place an infrastructure that provides IT as a service or a number of services. The services are made up from all aspects of IT, servers, storage and networking and need to be able to scale dynamically based on the real-time fluctuations in demand. With that in mind, we can define a number of points on utility computing:

- it delivers IT as a number of services;
- it provides the *right* services in the *right* amounts at the *right* time;
- it raises accountability for responsible consumption and reliable service delivery;
- it optimizes reliability, performance and total cost of ownership (TCO);
- it responds quickly to changing business requirements.

The first point here is very important, 'it delivers IT as a *number* of services', not a single service. This enables the ITO to look at its environment and its organization and decide which service could and should be delivered first. Later in this book we discuss how to pick that first service and make it successful.

The road to creating a service from IT (see Figure 2.4) is basically the same, whether it is backup or storage, servers or applications.

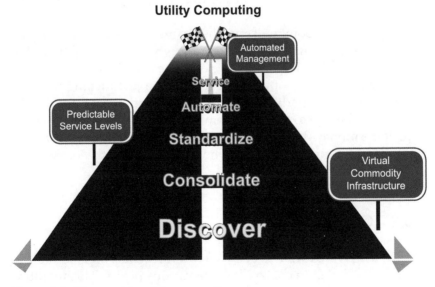

Figure 2.4 The road to utility computing.

More detail is given later in this book, but in essence, there are five steps:

1. Discover.
2. Consolidate.
3. Standardize.
4. Automate.
5. Create a true service.

One of the other benefits with utility computing is that while there is a road, it is possible to stop at any point along it and still realize benefits.

2.3.1 Discover

The last time businesses carried out a comprehensive exercise to discover what it was they had in their IT environments was in 1999, in the lead up to Y2K. Without a good picture of what is in the environment and how it is being used, planning to deliver a service effectively is impossible. Although time consuming, the effort put in to the discovery phase will reap dividends when services are being designed. A number of forms have been developed to help in documenting the discovery phase, and examples of these are given in Appendix B.

Stopping at this point will give the business a clear view of what it has, how it is being used and, as for Y2K, what is not being used. For servers and applications that are no longer required, maintenance and support can be dropped, not to mention just turning them off to save power, air conditioning, floor space and, most importantly, the time of the ITO to manage them. It behooves the ITO to make a note of the money that is saved, or can be potentially saved, during this part of the project, as it will help in justifying the project and others in the future.

2.3.2 Consolidate

Having discovered what is in the environment, it is then possible to put in place a plan to consolidate the various pieces. This might be done by examining what there is *most* of, what is easiest to manage, or the cheapest. It does not have to be a single option, in many cases enterprises want to consolidate onto a limited number of alternatives in order to keep their options open.

Stopping at this point enables the IT department to see the variety in its environment and help put in place plans or guidelines for best practice. It can also help in defining a purchasing strategy moving forwards, simplifying the environment.

2.3.3 Standardize

Standardization goes hand-in-hand with consolidation. In many cases, the order of these two steps is debated, do you standardize before you consolidate, or after? The point here is that consolidation will drive the standardization. When consolidating, it automatically creates standards that can then be put in place going forwards.

Stopping here will enable the ITO to put in place best practice around known *good* configurations. Purchasing will be simplified, as will management, as IT becomes a known quantity. Servers can be bought to a standard configuration, storage configured in a well-known way, applications installed and deployed in exactly the same way on every system. Because the configurations are known, the service level they provide, all things being equal, will also be known. Predictable deployment with predictable results, results in a lower TCO as management costs are reduced and by ordering template solutions from vendors, prices can be negotiated more easily. Vendors can also hold spares as they know which machines will be required, these may even be held on the customer's site to further reduce downtime.

2.3.4 Automate

The next step on the road to becoming a utility is automation. Standardization is not a prerequisite to automation, although it certainly makes it easier. Automation frees up time for valuable IT staff by removing the drudgery from many of the tasks an ITO carries out, and enables consistency of operation. While this can be supported with tools, the place to start is with a simple written process. What are the steps an administrator takes in order to ensure a new server is backed up properly? What happens when a new disk array arrives at the loading dock before it can be used? What happens when a database needs to be created? Automation is a scary idea and many shy away from it, believing that they will lose control. What is meant here by automation is that best practice has been formalized and, in some cases, codified in tools. Each step in the process might be fully automatic, or it could be semi-automatic or even completely manual. When a disk array arrives on the dock, perhaps the first step is to 'check for damage', obviously this does not happen automatically, but perhaps there is a stage later in the process, 'add to asset management system', which can be fully automated.

Stopping at automation, perhaps with best practice written up, or maybe with some simple tools such as process automation workflows to help consistency of deployment, will reduce costs. Enabling lower-skilled personnel such as operators to carry out tasks that previously required higher-skilled administrators is an obvious cost reduction but more importantly, it will allow the IT environment to scale because of the consistency.

2.3.5   Service

The final step along the road is to turn the IT function offered into a true service. Monitoring of the service to ensure that the pre-agreed service levels are met and visibility into costs based on usage, or perhaps some form of internal chargeback, are critical in evaluating whether the investment in IT is delivering a proportionate amount of business value.

Stopping with service, isn't this the end of the road? Well it is the end of one road but there are still a couple of items that must not be forgotten. The first is innovation, without innovation the service will die. Would you still sign up for a cell phone service that did not have voice mail? Innovation is supposed to increase because of utility computing, so forgetting to do it now would be unfortunate! The second item is decommissioning services. In many cases, this is not a complete decommission of the service, but probably a replacement of one service with another. For example, a low-end server might no longer be available as it has been removed from the manufacturer's catalog. The ITO then has the decision as to whether to replace the server in the *standard* configuration with a new one, or to create a new service around it, maybe with the same name. Either way, the service and the associated cost will change.

There are two other items that are core to an operational IT utility, one we have already alluded to, automation, the other is the service level agreement, or SLA.

2.3.6 Service level agreements

Utility computing focuses around services and the associated service level agreements (SLAs). These are the contracts between the ITOs and the consumers of IT, the lines of business. There might be one SLA or many. In essence, each is a contract, much like that which exists between a home owner and the traditional gas, water or electricity utilities. It specifies what the consumer is to expect, in a language that they understand, the costs associated with using the service, along with any penalties that might be payable should the service not perform as expected.

Several businesses that are in the early stages of implementing a utility computing strategy are not keen on the word 'agreement' and so use 'expectation' instead. This tends to be exactly the same but does not have any financial penalties associated with not meeting the service levels.

SLAs are made up of a number of pieces; the most important thing is that they can be monitored. Without objective measurements, they are not worth the paper they are written on. Often, an SLA will be made up of a number of constituent parts called service level objectives (SLOs). These have specific attached metrics, for example, data will be backed up every night.

As an SLA is between the ITO and the line of business, it is important that it is written in language that is understood by both parties. Talking in terms of i/o rates may not be appropriate, whereas transactions per second could well be language of the application owner. Terminology, or rather the interpretation of terminology, is often an issue when putting SLAs in place and is one area where the ITO needs to adapt to learn about the lines of business in order to interact with them efficiently.

From the perspective of the ITO, the only realistic way to deliver an SLA is to deal in known quantities. The creation of service templates, which, in effect, are known good configurations is the most effective way this can be done. These templates can then be costed effectively from both capital and management perspectives. By creating multiple templates for the same service, but which offer different service levels, IT can be delivered to each area of the business at an appropriate cost. Creating services and SLAs is often the 'showstopper' for many businesses and a practical approach is given in Chapter 5.

Templates drive standardization, which, in turn, enables automation to be carried out more easily.

2.3.7 Automation

In order to meet SLAs and to increase efficiency and effectiveness, automation tools need to be adopted. It is curious that while automation of business processes has long been a way to improve business efficiency, and that IT has been the technology to support it, it has never been carried out on IT itself. There are a number of reasons that this has been the case. First, it was thought that IT was the domain of an expert whose knowledge could not be improved upon. Secondly, that there were not enough 'standard' operations that needed to be carried out on a regular basis, and thirdly, that the tools to support such tasks just did not exist.

Automation of IT falls into two categories:

1. workflow;
2. policy.

Workflow within the IT domain is subtly different from that which usually springs to mind. The concept is the same, there is a task to be done and it has been broken down into a number of steps. The steps are generally carried out sequentially, although some will be done in parallel, and they can be completed by one or more different people or roles within the business. Traditional workflow is a series of operations that occurs between various individuals or groups, often around the creation and distribution of documents. Within IT, it is a series of operations that occur within the environment. For example provisioning a new server or running a backup task. Also, within an IT-based workflow, the number of steps is often few and it tends to be a sequential set of steps with very few branches.

Workflows tend to be created by capturing best practice, codifying them with tools and then enabling other, less-skilled personnel, for example operators rather than administrators, to govern them.

Automation through policy is the second area where efficiencies can be improved. Unfortunately, policy is an over-used word and means different things to different people. Taking the simplest definition for now, a policy is something that affects the behavior of a system. It is configured by the administrator and can be applied to similar types of objects, as well as to individual objects. Often, policy is a double edged sword, on the one hand, the complete control and automation of every aspect of an environment is a good idea, true 'lights-out' operation. On the other, the thought of configuring and maintaining the policies to do this is completely impractical. A happy medium needs to be found where policies can be set to carry out automatically the most common and tedious of tasks, without becoming an administrative burden in its own right. A simple example might be to grow a file system automatically *before* it runs out of space.

2.4   RETURN ON INVESTMENT

Utility computing is all about money. Money saved through better use of resources, saved through simplifying management and increased through being better able to respond to the lines of business. The ultimate goal is for the business to make more money. Effective use of technology is a business differentiator, utility computing as a process and an architecture can offer the competitive edge (Figure 2.5).

Figure 2.5 The business of utility computing.

Before starting on a utility computing project, it is important to have business metrics for the existing environment in order to tell whether the new services offered by the ITO are actually better. The sections that follow give some examples.

2.4.1 Hard metrics

- Deferred capital costs for storage and server infrastructure. By improving utilization and exploiting virtualization technology, existing resources can be used in preference to purchasing new ones.
- Reduced cost per TB or per server due to increased discipline of storage/server classification. Template solutions coupled with workflow automation reduce management cost.
- Reduced number of administrators/experts per TB or per server. Once again, automation is key to improving scalability within the ITO.

Although hard metrics, with objective goals, are important to measure in order to prove the return on investment, it is the subjective soft metrics that will be important for subsequent delivery of IT as a service.

2.4.2 Soft metrics

- Accelerated service time for new capacity requests. This can be measured through the request (or ticketing) system as the time from when the request is created to when it is satisfied. On a broader level, it can be measured as a more rapid time to market for the product that the business unit who is consuming the IT is delivering.
- Increased uptime due to better configuration decisions and dynamic configuration capabilities. Known good solutions based on templates ensure that problematic configurations are a thing of the past.
- Increased capacity for new service initiatives.
- Liability protection. Automation and best practice ensure simple tasks are not overlooked. For example, automating the addition of a new server or storage can include automating the protection of the data on the server.

2.5   WHY NOW?

Why is utility computing happening now? It is a simple matter of economics, the hazy days of 1999 and early 2000 before the .com crash have long gone. The ITO is now under the very close scrutiny from the entire business. Being a major cost center with no proof as to the value that is provided to the business is a distant memory and a comprehensive analysis of where the money goes is often taken (see Figure 2.6) to identify areas where money can be saved. It is now up to the ITO to prove its usefulness, if it does not or can not, then it will ultimately be outsourced.

2.5.1   The relationship to outsourcing

From 30 000 feet, utility computing and outsourcing share a lot of characteristics:

- IT is delivered as a service;
- you pay for what you want;
- you pay for what you use.

However, there is a key difference – penalties. Many outsource deals are done for 5–10 years and terms are based on current and planned usage. They do not take into account breakthrough technologies or even breakthrough new business. Should either of these occur, the business pays a penalty. In fact it pays in several ways. The

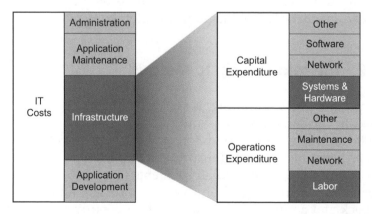

Figure 2.6 Typical IT expenditure profile.

breakthrough technology might reduce costs of IT significantly, for example, the introduction of serial advanced technology attachment (SATA) disk drives both drove down the cost of storage and made way for new ways of doing things, most commonly the move away from tape for backup. For the outsourcing organization, they could save money by utilizing SATA drives as well as offering new services (at increased cost) relating to improved backup recovery times. The outsourced organization saw none of these benefits without increasing the cost, as any margin derived from the change was held on to by the outsourcer.

New breakthrough initiatives within the business also suffer as the technology they require to support them might not be part of the original deal, resulting in a renegotiation or, worse still, the creation of an independent IT capability and/or parallel service agreements. An extreme example here was the introduction of picture messaging for mobile phones. While the short message service (SMS) had been around for a while and was becoming well used in Europe, the average message length was only a few bytes. The introduction of cameras on phones resulted in messages of several hundred bytes per message, requiring several orders of magnitude of storage. For those companies that were in an outsource agreement, the increase in capacity requirements and the financial penalties they would have to pay to meet the new requirements would have seriously impeded the adoption of the service, as the costs would have been too great. Fortunately, mobile phone companies were able to react to the new technology in a cost-effective manner and picture messaging has now become commonplace. Of course, technology has moved on again and video messaging is on the rise, requiring another several orders of magnitude of storage.

2.5.2   The relationship to the grid

The grid (or Grid Computing) is another topic which is often mentioned in the same breath as utility computing. This time, from 30 000 feet the two are very different. Applications in the grid are seen as being niche parallel distributed applications, in which the data is processed in individual segments (c.f. batch jobs). Data is distributed out to the grid of servers, processed, and the results fed back and accumulated. Data is not necessarily required to be processed sequentially, and if one batch job fails, then it can always be run again at a different time. While the niches are big, for example, pharmaceutical drug discovery, geological survey analysis, financial risk analysis, they are not really seen as everyday applications. Most applications found in a business are expected to be running all the time, not as batch operations, and the sequence in which the data is processed is important. However, looking more closely, say from ten feet, the infrastructure required for a grid is very similar to that required for utility computing:

- What resources do I have?
- How are they being used?
- Can I do more with this resource?
- What is the cost of using this resource?

The answers to these questions in a grid scenario could result in new jobs being scheduled on the resource, while the results in a utility computing scenario could mean improved workload balancing, server or storage consolidation, roll out of new applications onto existing hardware and, of course, the ability for chargeback.

2.6   IT IS NOT ALL OR NOTHING

Successful implementation of a utility computing strategy would result in a very dynamic ITO that could respond instantly to new business requirements. While this book looks at how the entire ITO could become service-driven, it is important to note that this is not an all-or-nothing approach. There are many services that can be provided in a utility manner, ranging from backup to storage to servers, and from databases to applications to disaster recovery and other higher-level services.

The timeline for creating the ultimate global IT utility (see Figure 2.7) gives an indication of when various IT functions either have been, or will be, delivered as services. The steps along the path from a technology perspective are defined by the maturity of management tools, the business applications available and the business model.

The details of where to start assessing your organization's readiness for the utility computing model are covered later in Part Two,

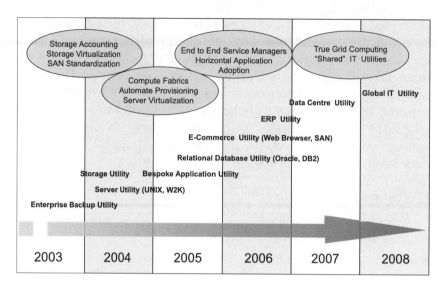

Figure 2.7 Timeline for utility computing.

but it is worth initially considering the various 'obvious' options. Where would be a good place to start an initiative?

- development;
- test;
- production services;
- offline storage/backup;
- online storage;
- servers;
- applications;
- other higher-level services, for example disaster recovery.

Starting with storage (see Figure 2.8) is an obvious choice for a number of reasons.

2.6.1 Offline storage/backup

In many cases backup is the obvious place to start. Businesses, or more importantly the individual lines of business within, accept that backup is a specialist function. They did not want to manage tapes

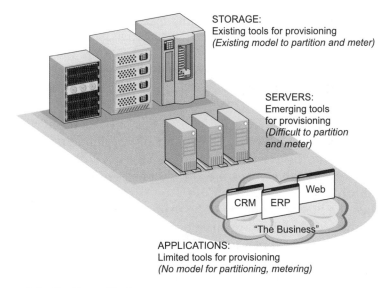

STORAGE:
Existing tools for provisioning
(Existing model to partition and meter)

SERVERS:
Emerging tools
for provisioning
*(Difficult to partition
and meter)*

Web

CRM   ERP

"The Business"

APPLICATIONS:
Limited tools for provisioning
(No model for partitioning, metering)

Figure 2.8   Starting with storage.

and did not want to purchase tape drives, let alone tape robotics, so it has become a centralized function that the ITO carries out. Add to this the fact that the tools for managing, provisioning and metering backup are mature. However, one of the questions often asked is that backup already is a service, so why change it? The answer lies in the visibility. As everyone is currently using it as a service, and often it is fully automated, the additional reporting for usage is comparatively small to implement. Costs associated with backup are often thought of as being those relating to the amount of data backed up. While this is true, it is not the whole story, the complete cost should include the media that is stored with old copies of data and how much data is restored. Restoring data often involves manual intervention and so pushes the cost up. Just the insight this type of information brings is often enough to make lines of business think differently.

The other reason to choose backup is that it can be introduced as a service in all environments, from development to production with minimal business risk – as long as the backups happen. Transitioning to template server solutions in a production environment obviously carries more business risk, along with the greater rewards. Development of process for service introduction, no matter what the service is, is essential, and resolving teething issues while minimizing risk is clearly good.

2.6.2   Online storage

Online storage is one area which has not only seen phenomenal growth over the last few years, with no end in sight, but it has also seen a huge increase in personnel to manage it. Often, complexities in the environment have resulted in specialist skills and the creation of dedicated roles.

While the business might not be ready to retrofit a utility model to existing storage, it is a great opportunity for new storage that is being bought (see Appendix A for a case study). As with backup, mature tools exist to monitor and manage storage efficiently. Virtualization technologies exist to enable the business to move from one hardware vendor to another seamlessly and without impacting the application. This is particularly important when considering the future of the service. Innovations, either technologically or from a business process perspective, might mean a wholesale change from one type of online storage to another; without virtualization, there will be an impact to the business, with it, the ITO is free to carry out the changes as and when it wants, and providing it maintains the SLAs to which it has agreed.

2.6.3   Servers

Server utilization has always been a subject of contention. For a long time it has been known that server utilization, like storage utilization, has been less than 30% (Bittman, 2003)[1] and therefore the promise of improved utilization that utility computing offers appears a natural fit. However, it is only in the last 12 months or so that the tools for provisioning and monitoring have started to emerge. New architectures, such as those promised by blade computing with their promise of an agile repurposable environment, also seem to match well with the utility computing story. New virtualization technology means that everything from the server

---

[1] Mainframe utilization is approximately 80%, RISC servers are approximately 40% and Intel servers are between 10–15%.

itself down to applications can be virtualized; this offers the ITO the possibility to increase physical server utilization by moving virtual servers and/or applications appropriately.

While utility computing can be applied to new server purchase and deployment, it can equally well be applied to existing resources. A "discovery" phase will have to be carried out on the existing resources before a project starts. When this is carried out, it is worth noting information such as resource utilization, as this will help with a consolidation process.

The biggest challenge in putting in place a server utility, especially one using existing hardware, is in overcoming organizational culture. With the exception of the mainframe, computer resources have traditionally been owned by the lines of business whose applications run on them. The servers were specified by the business and, once purchased, given to the ITO to run and maintain; no other applications, especially not those from different lines of business, would be run on them. In the utility world, this will no longer be the modus operandi, the ITO will own the servers. The servers deployed will be based on templated known good configurations and the ITO will be, in theory, able to replace them at any point in time – providing that the agreed service levels are still met.

While servers are not an obvious place to start, if a business is considering a new server farm, then being able to apply utility computing principles to it, to enable a more flexible and efficient environment, is a worthwhile investment.

2.6.4 Applications

The bespoke nature of applications and the current state of tools to provision them make turning application delivery into a service extremely hard. However, there are exceptions. Specific applications, for example databases, are often configured in exactly the same way and so can be considered suitable for delivery as a service. Other applications, such as web servers or scalable front-end application processors, are also suitable. In these last two instances, it is often a requirement that processing be able to be switched on or off on demand. In this case, a utility architecture based on templated server and storage configurations is ideal. Automation can be used in the provisioning and reprovisioning of both the servers and the applications, based on demand.

2.6.5 Other services

Another service that can be considered for delivery in a utility environment is disaster recovery (DR).[2] While this is very specific to an individual business, it is possible for a business to define the various levels of service it needs from a DR service. This takes on all the classical aspects of delivering IT as a service, talking to the consumers about their requirements, defining a limited number of service levels and ensuring that *all* the business applications fit into them and then creating an implementation plan. The difference between applying utility computing principles to DR and traditional DR is the creation of a limited number of options, thereby decreasing the cost associated with delivering the service. In one particular customer's case (see Appendix A for the DR case study), this resulted in requiring half the initial estimated budget.

Any 'service' that the ITO provides to more than one business unit or line of business should be considered for delivery as a utility-based service. For those bespoke services that are only used by a single consumer, it is worth spending the time looking closely at why this is a special case. Is there a good business reason, or is it that they do not want to change? Cultural change across the whole organization will be required if utility computing is to be a success. If there is a good business case, understanding the actual usage and being able to attribute costs to providing that individual service will be essential if the ITO is to prove its true value to the company. Over time, if costs are hidden from the CIO, then the true worth will not be seen.

It is up to the business to decide which is the best for them and then make a success of that before moving on to other, more complex services. Even within these areas, further partitioning can occur, for example, applying utility computing principles to a new blade server farm or new disk storage farm. While it is very possible to start using a 'brown field' site, it is often easier, especially for a proof-of-concept, to start with a 'green field' opportunity.

---

[2] Disaster recovery is used here rather than business continuity as we are talking specifically about the IT aspects of a business continuity initiative, which has significantly greater scope, encompassing all the resources, such as people and buildings, as well as IT.

2.7   FURTHER IMPLICATIONS OF UTILITY COMPUTING

While the high-level implications of improved efficiency, effectiveness and responsiveness should be clear, the more far-reaching implications are those within other areas of the business.

Currently, most organizations are split into different lines of business, each one having its own applications and, more importantly, its own hardware on which the application is run. Typically, the hardware is specified by the line of business, bought from its own budget and then handed over to the ITO to run and maintain. The hardware is seen as belonging to them – and no-one else will touch it. For utility computing to really succeed, this attitude needs to change. In the same way that tape libraries are shared between departments for cost purposes, servers and storage need to be handled the same way. Purchasing decisions should be driven by the ITO having listened to the lines of business, rather than the other way around. The lines of business need to have agreements with the ITO on the services they are providing and the levels to which they are provided. This will give the ITO the opportunity to save costs by purchasing in bulk, save management costs by using 'standard' configurations, improve efficiency through server consolidation and improve effectiveness by utilizing standard templated hardware and software stacks that can be quickly rolled out to meet business needs.

With the ITO fully in charge of resources, they can make best use of new technology and, providing they keep to the previously agreed SLAs, there is no reason why they could not change the entire infrastructure beneath the applications if it made sense. The cultural shift to not owning any hardware and running on SLAs will be a big but very strategic one and will be discussed later in this book.

Of course, the change to a service-driven organization means the introduction of new roles, such as the Customer Relationship Manager and even a Services Manager. These people will be responsible for finding out what the consumers of the IT services offered think, looking for improvements in the services, communicating planned outages to pre-empt support calls and fixing problems that occur. The role is not new within a business, but it is new within an ITO.

For maximum efficiency and effectiveness, suppliers also need to be aware of the change in culture of the organization. They too need to respond to requests in a timely manner. The negotiation of service and supply contracts will have to be done bearing in mind that the supply may need to go down as well as up. It is not practical to think that the ITO will always have the spare capacity to meet every need from every line of business; it will be up to the suppliers also to manage some of that capacity and, therefore, risk.

2.8   SUMMARY

- The utility computing message is simple – costs down, efficiency up. The hard part is proving that is the case.
- The road to utility computing, discovery, consolidation, standard-ization, automation and, finally, true service delivery, is the same, no matter which IT service is being considered.
- Service level agreements tie IT to the business and are vitally im-portant in the next generation of IT organization.
- Automation, best practice and template solutions drive down cost and drive up efficiency and agility.
- Utility computing shares ideas and concepts with outsourcing, but ensures that the business remains in charge of its own IT destiny.

REFERENCE

Bittman, T. (2003) *Predicts 2004: Server Virtualization Evolves Rapidly.* Gartner Research.

Saxe, J. G. (1964) *The Blind Men and the Elephant.* World's Work.

3

Historical Trends, or 'Is Utility Computing Really New?'

3.1 OVERVIEW

Arguably, utility computing is the original form of computing. This chapter looks at the roots of utility computing and finds they are not as different from early computing models as expected. It also traces some of the technological advances that have occurred to bring us to this point.

3.2 BACK TO THE BEGINNING

The high cost of early mainframe computers made them inherently large business resources. The computer (usually singular) was shared out of fiscal necessity. The technology of the time also encouraged utility-style operation. Ordinary people did not interact with computers. They prepared 'jobs' on punched cards, which they submitted to specialists who, in turn, queued them for execution in an order determined by the user's importance. As jobs completed, printed results would be returned to their submitters.

Delivering Utility Computing.   Guy Bunker and Darren Thomson
© 2006 VERITAS Software Corporation. All rights reserved.

During the mainframe era, it was common for IT departments to charge users for the system resources they consumed. Charges would be based on resources used, with rates reflecting relative priority and other factors, such as time of day. In effect, a user's willingness to pay would determine their priority (level of service).

In the mainframe era, chargeback was a fiscal necessity because of the high cost of computing; it was just too expensive for detailed cost accounting to be ignored. Moreover, cost accounting was technically easy: a user's job would run on one computer, whose utilization could be measured; data was stored typically on removable 'disk packs,' each allocated to a single application; printers and other resources were concentrated in one place and their use was associated with jobs easily.

Thus, computing in the mainframe era operated very much as a utility. Services were limited, delivery was centralized and there was strict accounting and billing. The utility was an exclusive one, however. Because of the cost, computing was only available to large corporations, government agencies and academic institutions. Individual departments, small users and the general public had no access.

3.2.1   Open systems and the utility model

During the 1960s and 1970s technology advances steadily delivered smaller and less expensive computers (Figure 3.1). In the

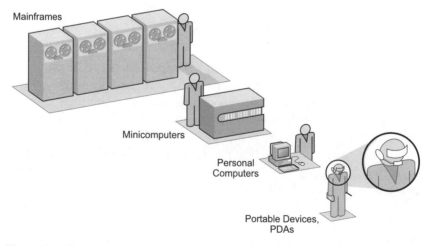

Figure 3.1   The evolution of business computing.

1980s, UNIX-based open systems matured, making administrative skills transferable. Computer usage increased accordingly, and servers dedicated to individual business functions, like order processing and inventory control became common, even in small businesses.

Paradoxically, the commoditization of computing diminished the utility delivery model, as single-purpose low-cost computers transformed computing from a shared service into a departmental expense item. Computing went from being delivered by a central IT department to being delivered to each department by departmentally-owned computers. Computing came to be treated as departmental capital outlay, rather than as an ongoing expense shared across the business. Sharing was limited in any case as computer platforms were proprietary and, even within a single vendor's products, configurations were specific to the application that was going to be run on them.

3.2.2   Personal computers

In the 1980s, computing became available directly to consumers, starting with kits, and followed closely by complete self-contained personal computers (Figure 3.1). Over the course of the decade, prices plummeted and the number of users and applications rose correspondingly. Individuals could afford computers that could do things they considered useful – writing letters, keeping budgets, playing games and exchanging electronic mail. It did not take long for these low-cost devices to find their way back into business, supplanting 'dumb' terminals connected to large mainframe servers.

The usefulness of a personal computer increases with its processing power. Today's personal computers are as powerful as the supercomputers of 1990, and easily capable of running applications that once required large data centers. Computing power has effectively become a commodity due to standards – it is low in cost, and deliverable in interchangeable units by a variety of vendors.

But low-cost processing by itself would not have been sufficient to engender utility computing; a distribution network was also required. The utility computing concept described in preceding chapters would not have been possible without a parallel development in information technology – Ethernet – and the pervasive networking that resulted.

3.3   CONNECTIVITY: THE GREAT ENABLER

From the earliest days of computing, it was clear that interconnecting computers would increase their usefulness dramatically. The defense establishment acted on this belief early – in the 1960s, the RAND Corporation sought ways for the United States to maintain control of its missiles and defense systems after a nuclear attack. The research led to the development of packet-switching network protocols with very flexible routing and a high degree of robustness.

3.3.1   Peer networks

The United States government's Advanced Research Projects Agency (ARPA) used packet-switching technology to construct the peer-based ARPANET. The ARPANET initially linked universities; later it was extended to include defense contractors. Although computers had been interconnected before, the ARPANET was the first one designed for peer relationships among computers – instead of one master computer issuing commands to slaves, any computer might make requests for service, and these might be granted or denied by other computers.

From ARPANET beginnings, peer networking technology evolved into TCP/IP protocol standards, and thence to commercial networks, as both users and developers realized that with standards for interconnecting computers, new ways of doing business would be possible. Low-cost computing could be carried out close to users and data sources, without sacrificing the control inherent in a central data center. Even more radically, with applications like electronic mail, peer relationships could be extended all the way to the personal computer level.

3.3.2   Ethernet

Ethernet was developed in the early 1970s to provide low-cost, high-performance interconnection for hundreds of computers in a data center or on a campus. Its development followed roughly the same path as TCP/IP – initial development by consortium, commercial introduction, market success and standardization. Combining Ethernet with TCP/IP made it possible for widely separated applications to interact, even across platform boundaries. Thousands of Ethernet local area networks were linked into both business and public network fabrics, and the age of client–server computing began. In the 1980s, government, academic and commercial businesses created backbone networks, both to integrate their own operations and to cooperate electronically with other businesses.

With Ethernet and TCP/IP, businesses could deploy low-cost personal computers and small servers anywhere. Users were no longer dependent on a central system; they could process data even when not connected. As a result, both processing and data became distributed throughout businesses. The utility aspect of business computing was diminished further by the relative independence of many departmental and individual computers treated almost as private property by their owners.

Just as Moore's Law predicts increased processing speeds, Gilder's Law predicts increases in network capacity. In 1995, George Gilder postulated that deployed network bandwidth would triple annually for the next 25 years and that individual link bandwidth would double every nine months. Developments to date suggest that this prediction will be accurate (See Figure 3.2), wireless access is growing with hotels and coffee shops providing affordable, and often free, connectivity, not just for business users, but also for the masses. Even the multitude of devices that people carry are being connected together through personal area networks (PANs), such as Bluetooth.

The elements for a return to utility computing are thus in place – a low-cost commodity (computing power) to deliver, and a means of delivering it (ubiquitous networking). All that stands in the way of a return to utility computing is a value proposition (the why) and a catalyst (the how).

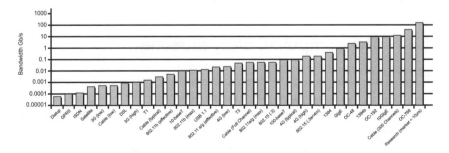

Figure 3.2 Telecommunications bandwidth numbers.

3.4   THE WEB AND THE RETURN TO UTILITY COMPUTING

In 1990, Tim Berners-Lee of CERN (the particle physics laboratory based in Geneva, Switzerland), presented a momentous paper describing a hypertext-linked system for information sharing within the high-energy physics community. The attraction was immediate and events followed rapidly. A year later, in 1991, the first version of the Hypertext Transfer Protocol (HTTP) for transferring text between computers of different architectures running different operating systems appeared. This set the foundation for today's World Wide Web.

In 1992, CERN opened access to the World Wide Web to the general public. The number of attached computers grew from 1.1 million in 1992 to more than 6.5 million in 1995. Today, hundreds of millions of computers and other devices are connected to the Internet. Televisions, games consoles, personal digital assistants and other devices use the Internet to share and find information. Text has given way to images, audio and video. Live television feeds, real-time multiplayer games and home shopping are all common. Never has so much information been available to so many people so simply. The web has moved computing from the domain of business applications to universal availability and utility to everyone. Grandparents, two-year-olds and everyone in between can enrich their lives through the use of the Internet.

Today, mobile telephones and personal digital assistants use General Packet Radio Service (GPRS) to maintain a continuous flow of information with the Internet. Just as the mobile telephone made voice connectivity independent of an individual's location, the Internet is on the way to placing any and all information at hand, no matter where one is or what the time.

3.4.1   The computing utility redefined

The Internet has taken the computing utility concept to a new level by virtualizing information resources. Today, anyone can use the Internet to find information on virtually any topic, without knowing anything about either the infrastructure or the source of the information. It is easy for a business to set up a website and conduct business online, and for users to buy in the comfort of the home.

Observing the rapid success of the Internet, many businesses have found ways to exploit the concept for internal use. Browser-based user interfaces obviate the need to install software on every client computer and train users for every new application. The cost savings from not having to distribute, maintain and update every application on every desktop computer are quite remarkable.

3.4.2   Changes in computer systems

The Internet has made delivery of information ripe for a utility. Information is there for the taking. But a utility must be always on; power must be there when an appliance is plugged in. Computers cannot always maintain this level of service – they fail.

From the mid-1990s, mission-critical information services have been protected from computer failure by Failover Management Software (FMS). This monitors continuously and directs applications to migrate or *fail over* to another computer when a failure is detected. Today, FMS can pool dozens of computers into *clusters*. Applications that run on clusters can move or redistribute themselves if a computer fails, without involving users, who are generally unaware of the specific computer with which they are interacting. Applications are *virtualized*.[1]

All the building blocks for utility computing can be found in today's data center: low-cost processing and storage, the delivery network, a universal user interface (the web) for anywhere anytime delivery and user motivation (the value of information and capabilities available on the web to both the business and individuals). Even the attitudes are in place – Internet users are supremely indifferent to details – they just expect the service to be there when they need it. It is worth noting that this expectation is also driving demands for service levels, which some organizations cannot afford to deliver.

3.4.3   Changes in Applications

The change technology, especially that brought about by the web, is nothing compared to the changes in applications and what is now

---

[1] The term 'virtualization' is used in many contexts in information technology. It is a layer of software between an application and the computers it runs on that insulates the application from hardware characteristics. A virtualized application can be moved from one server to another easily without user awareness.

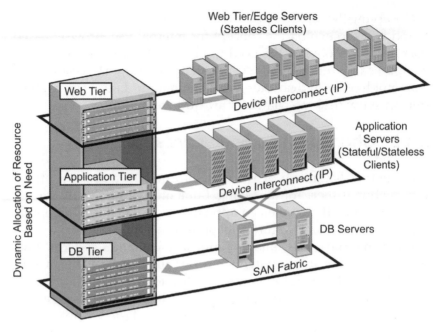

Figure 3.3 Application complexity means three tiers (or more) are the norm.

expected by consumers. This also affects what they expect from applications they use when at work. People expect information to be available at all times, applications to be responsive. New *N*-tier architectures (Figure 3.3) have increased complexity, and data presented to the consumer can come from multiple sources. Inconsistent demand for processing power due to daily-or even seasonal, requests has made it harder and harder for the ITO to use the resources they have efficiently. Applications have to be designed and built with peak demand in mind, predicting demand three years hence can be akin to crystal ball gazing, with the result that a lot of IT resource is underutilized and even those that are well used at peak time have poor average usage.

3.5 PAY-AS-YOU-GROW DATA PROCESSING

The 1990s saw a change in the way business computers and storage were designed and delivered. Users looked for hedge strategies against unanticipated growth, and vendors began to design large frames that could be purchased sparsely populated with disk drives, processors, memory, and so forth. As capacity needs grew, additional components could be added to these frames with little disruption. In software, capacity-based licensing came into being – the purchase of a license for a specific number of users. In both cases, users would purchase a 'big box' for a relatively low price, and add capacity as requirements increased. Incremental capacity might be unlocked with a software license key, or delivered as add-on hardware components.

This was progress, because it allowed businesses to pay only for the resources they required. The drawback was that it bound businesses to a single supplier. To take advantage of new or superior technology from other suppliers, they would have to write off the big box investment.

Software companies have started to move further – to usage-based pricing.[2] Attempts to introduce it in the year 2000 failed, but with the slower economy since that time, the value proposition for users is increasingly attractive, and pressure on software companies to adopt the model continues. Usage-based pricing would allow, for example, a business that required a database for a one- or two-month project to buy a license just for the duration of the project. In most cases, reducing license cost would also reduce periodic maintenance charges, with further cost savings to users. A move to more flexible licensing by software and hardware vendors would drive utility computing, from a cost/pay-as-you-go perspective, more rapidly into the mainstream; its lack remains one of the greatest hindrances.

3.6 UTILITY COMPUTING AND THE INDUSTRY

Major hardware and software vendors have realized the attraction of utility computing to users. Even the business analysts are in on the act and have their own variants, with the result that the market projections vary wildly from $2B to $70B by 2006. For the hardware and software vendors, most have made it a significant technology

---

[2] Usage-based pricing is key to utility computing. With usage-based pricing, fees are charged based on actual days of usage. This model poses a challenge for software companies – how can they maintain revenue streams and at the same time save their customers money. Typically, daily prices tend to be higher than annual licenses, but for occasional use, usage-based pricing can be advantageous.

and marketing theme. Companies promoting utility computing initiatives include:

- BMC (Business Service Management) – http://www.bmc.com;
- CA (On-demand Computing) – http://www.ca.com;
- Cisco (Business Ready Data Center) – http://www.cisco.com;
- EDS (Agile Enterprise) – http://www.eds.com/services/
  agileenterprise;
- EMC (Auto IS) – http://www.emc.com;
- Forrester (Organic IT) – http://www.forrester.com;
- Fujitsu (TRIOLE) – http://www.fujitsu.com/global/services/
  solutions/triole;
- Gartner (Policy-based Computing) – http://www.gartner.com;
- Hewlett-Packard (Adaptive Enterprise and Utility Data Center) –
  http://www.hp.com/large/infrastructure/utilitydata/overview;
- IBM (Autonomic Computing) – http://www.research.ibm.com/
  autonomic;
- Infoworld (Dynamic Enterprise) – http://www.infoworld.com/
  infoworld/article/03/04/11/15dynamic_1.html;
- Mercury Interactive (Business Technology Optimization) –
  http://www.mercury.com;
- Microsoft (Dynamic Systems Initiative) – http://www.microsoft
  .com/presspass/press/2003/mar03/03-18dynamicsystemspr.asp;
- Oracle (Commercial Grid) – http://www.oracle.com/
  technologies/grid/index.html;
- SAP (Adaptive Computing) – http://www.sap.com;
- Sun Microsystems (N1) – http://www.sun.com/software/
  learnabout/n1;
- VERITAS Software Corp (Utility Computing) – http://www
  .veritas.com.

Arguably, IBM started the trend with its ambitious Autonomic
Computing initiative, whose goal is to create a self-managing
IT infrastructure of self-configuring, self-healing, self-optimizing
and self-protecting servers, networks and applications. Autonomic
computing has caused other software and hardware companies
to think differently about the architecture of future IT solutions,

but it will be some time before it becomes a commercially viable reality.

VERITAS's strategy is somewhat unique in being a multiplatform software-based strategy that specifically does not attempt to direct businesses towards common server, storage or network platforms. This strategy is particularly well-suited to larger businesses, as well as system integration companies, several of which are developing IT utility architecture service offerings. As a software-based strategy, it does not require wholesale replacement of existing equipment, and is, therefore, practical from an evolutionary standpoint. As a multi-platform strategy, it provides the system integrator with skills that can be transferred from client to client, no matter what computing and storage platforms the clients use.

Vendors' utility computing initiatives all have the same goal: automatic dynamic configuration of hardware and software resources based on proactive rule-based policies to improve utilization and reduce cost.

But the full realization of utility computing is more than lots of low-cost hardware and software from one vendor. Data centers today are inherently heterogeneous. Whether due to legacy or to deliberate decisions to run applications on the equipment best suited for them, most data centers have a variety of Linux, Windows and UNIX equipment in place. Utility computing should make at least some of these resources interchangeable – to allow, for example, excess storage capacity purchased for electronic mail to be co-opted as temporary storage for month-end closings. In a computing utility, the computing and storage resources should be genuine commodities that can be delivered effectively wherever they are needed.

3.6.1 Standards

In an ideal computing utility, applications would be able to find and interact with other applications, as well as processing, storage, databases and other resources supplied by different vendors. Resources would be able to interact, regardless of where they came from. The history of computing demonstrates that successful inter-action is achieved through standardization. Several utility-related interoperability standards are already in the early stages of user adoption:

- Web Services Description Language (WSDL);
- Universal Description Discovery and Integration (UDDI);
- Simple Object Access Protocol (SOAP);
- The Common Information Model (CIM);
- The Storage Management Interface Standard (SMI-S).

While they certainly represent progress, these standards actually compete with each other in some respects, and it will be some time before interoperability standards are mature. Users are beginning to adopt utility computing, despite the lack of mature standards. But this is actually encouraging – history shows that user adoption of computing technology drives standards, rather than the reverse.

Standards for application provisioning lie further in the future. Today, each system vendor offers its own platform-specific application-provisioning facilities. Some software products support multiple platforms. While better than homogeneous solutions, these are still not ideal. Without interoperability standards, users are limited to homogeneous environments, or at best, multiple 'utilities' – one for each vendor whose platforms are present in their environments.

Even for a single application, standards are necessary so that data can be processed by different platforms, for example, so that an Oracle database can be accessed by either Windows or Linux servers. Without application data format standardization, users will not be able to benefit fully from utility computing.

3.7 SUMMARY

- The basic concepts of utility computing – centrally provided common services, resource sharing and accountability – are as old as computing itself.
- The utility model fell out of use when low-cost, easy-to-use computers made it possible for users (business and operating units) to control their own data processing.
- Simultaneously, standardization in networking and availability of low-cost processing and storage, have set the stage for a re-emergence of utility computing.

- Other changes to the way in which software, and to some extent hardware, are purchased and used are contributing to the utility computing paradigm.
- Ultimately, standards will be needed for full realization of utility computing, from protocols allowing applications to intercommunicate, to management of the utility infrastructure. These will take time to mature, both technically and politically, but user adoption of the utility paradigm will hasten that maturation.

4

The Utility Model in Detail

4.1 OVERVIEW

The premise of this book is that information services can be delivered more effectively by delivering them as a utility, rather than by the ad hoc custom infrastructures in common use today. A good starting point for appreciating what an information technology utility is, is an understanding of what makes any company a utility. In this chapter we will explore briefly what essential attributes are associated typically with a service to make it a utility. We will then study a traditional utility service in some detail in order to compare its specific attributes and characteristics with the way in which we *could* deploy information services using a utility model.

The chapter will not discuss in detail how possible it is today to replicate the attributes of a traditional utility in the IT world. This will be discussed later.

It is important that, before diving into specific technical detail concerning utility enablement for IT, the reader first has an appreciation as to what utilities are really for (and what advantages they are capable of providing in a generic sense).

Delivering Utility Computing.   Guy Bunker and Darren Thomson
© 2006 VERITAS Software Corporation. All rights reserved.

4.2   THE ESSENTIALS OF A UTILITY

Traditionally, a utility has been defined as a business that performs essential public services[1] subject to government regulation.[2] While the utility business model, rather than its relationship to governments, is of interest in this book, this definition implies several characteristics that are relevant when applying the utility model to enterprise computing:

- a utility provides *essential* services;
- a utility provides a small set of well-defined services to a large user community;
- a utility is *reliable*;
- a utility provides *appropriate* levels (or 'classes') of service;
- a utility is measured (and differentiates itself) based on the *quality* of service that it delivers;
- a utility is run as *'a business'*.

As we will discover later, these attributes have to be explored carefully and understood fully by an organization that is considering the transformation of a service into a utility. This is particularly true of information services where, traditionally, heavy focus has been given to technical excellence and functional uniqueness, as opposed to quality of service and profitability.

As the utility model is explored, it is also worth noting that, traditionally, utilities have succeeded or failed (often quite spectacularly!) largely due to some, or all, of the attributes listed above.

4.3   THE UTILITY MODEL IN DETAIL (AN ANALOGY)

In order to explore fully how a utility works and what it is capable of providing, we have chosen in this chapter to use an analogy, relying on an area of service provision that we all encounter regularly in our lives ... a restaurant.

---

[1] A public service is the business of supplying a commodity (such as electricity or gas) or service (transportation) to any and all members of a community – *Webster's New Collegiate Dictionary*, 1979.

[2] Different countries regulate their utilities differently, even oscillating between public and private corporations at different times. It is generally true, however, that the commodities provided by utilities are regarded as sufficiently vital to be regulated as to price and quality of service.

The restaurant business is not often referred to, or regarded, as a utility. In writing this book, however, it seemed to us that a restaurant fulfills absolutely each of the criteria mentioned within the previous section:

- it provides *essential* services – we have to eat;
- it provides a small set of services to a large group – menus are designed to provide limited options and yet appeal to a broad set of people;
- it is *reliable* – we expect to get what we order in a timely fashion and without being poisoned;
- it provides *appropriate* levels of service – we typically select our restaurant based on taste and budget;
- it is measured (and differentiates itself) based on the *quality* of service that it delivers – poor quality of service has been known to put a restaurant out of business quicker than in many other parts of the service industry;
- it is run as *'a business'* – restaurants are run with profit very much in mind. (This is perhaps the biggest difference, as ITOs are seldom run with profit in mind, although cost reduction and improved utilization does result in increased bottom line profit for the organization overall.)

So, rather than relying on more obvious analogies (such as gas or water utilities), let us explore the utility model using a restaurant business as a template.

A utility service typically relies on a workflow that encompasses some fundamental processes, all of which are critical to the successful delivery of utility services. A generic set of processes could be described as follows:

- service planning and creation;
- service deployment;
- service provisioning;
- service assurance;
- service reporting and billing;
- service redefinition and retirement.

These processes are cyclical in nature and so represent a 'lifecycle' when applied to any utility service. Figure 4.1 illustrates how these various processes typically relate to one another and operate.

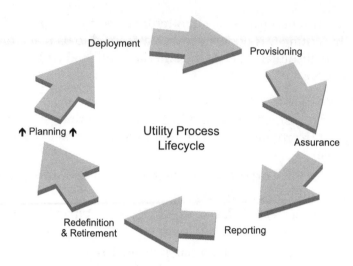

Figure 4.1 The utility lifecycle.

Let's take a look at each of the functions that exist within a generic utility workflow and apply them now to our restaurant analogy.

4.3.1   Service planning and creation

When a restaurant business is started, the new owner has to think long and hard about how to make their restaurant compelling to potential customers, different from the competition and economically viable in order for it to run (and grow) as a successful business (Figure 4.2). Market research has to be conducted in order to make

Figure 4.2 Success comes from extensive planning.

sure that the restaurant will be popular in its local area (local people are interested in its offerings and there are not lots of other, similar restaurants nearby, for example). All of this thought eventually ends up forming the definition of the service that the restaurant is going to provide.

Of course all restaurants provide the same basic service. All the more reason then, for a new restaurant to create service offerings that provide additional value or enjoyment to customers that cannot be obtained from any competitor.

In the utility world, a service provider must seek to provide consumers with service offerings that are truly differentiated. In other words, there is typically little point in approaching a market with offerings that can be obtained from elsewhere. History shows us that the consumers of utilities elect to make use of the utility service in order to make their lives easier. Once a utility service has been established and initially successful, utility service providers then start to compete with one another, largely using price and quality of service as competitive weapons. It is very difficult to entice a consumer of one utility to an alternative one, unless that alternative differentiates itself from its predecessor in a substantial way (almost always related to price or quality of service, which can include subjective things like ease of interacting with the support organization, or being able to carry out requests online rather than in person).

A note to IT strategists

Although everything stated so far may seem obvious when considering traditional utility services (or even a restaurant!), there are valuable lessons here for the IT department wishing to explore utility computing. Will the new IT service that you are creating be truly differentiated, and why? What will its competition look like over time?

If you do not believe you have competition, then it is worth organizing some interviews with your customers. Is the CIO considering outsourcing? Have any of your customers bought systems that they run themselves? 'Oh, you don't have to worry about that, it's just a little file server because we ran out of space and needed something quickly.'

So, during the service creation process, service definitions must be authored that describe customer offerings that are truly differentiated. Typically, service definitions are made up of several core elements, these include:

- a full description of the service proposition and function;
- classes ('tiers') of service definitions;
- service level agreement definitions (SLAs);
- cost of service definitions;
- basic service management and delivery workflows.

Let's go back to our restaurant analogy to explore how these various service definition elements can be applied to service delivery.

Service propositions

The restaurant's service proposition (sometimes referred to as the 'value proposition') will need, first of all, to be described in a concise and meaningful way, so that the business can describe (market) itself to a wide audience. In a restaurant, a service description often appears at the front of a menu so as to acquaint customers with the service that they can expect. Extracts from the service description will also be used in traditional advertising methods such as newspaper advertisements or media commercials.

IT utility marketing

Later in the book we will explore fully effective methods of advertising an IT utility's service proposition. In our experience, an organization's intranet site, as well as targeted email campaigns, can prove to be effective vehicles for advertising an IT service proposition to potential users.

Communication needs to be thorough, what are the benefits the users will get from adopting the new services? What is the relative cost to the way in which things are currently done? Are there going to be changes anyway? Communicating the effects before the changes occur, rather than afterwards, goes a long way towards building trust.

Classes of service definitions

In the creation of service propositions, it is vitally important to give users some choice in terms of the nature of the services to which they can subscribe. This may imply selecting a 'quality of service' that is appropriate or, in some cases (such as the restaurant), it may mean selecting service offerings that are appropriate to a need or desire.

The closest comparison to the definition of service 'classes', using our restaurant analogy, would be the creation of 'courses' that appear on the restaurant's menu. If the restaurant offers a three-course menu, then this gives its consumers choices that influence how much they eat, how much money they spend and how they combine the various offerings that are purchased so as to ensure a pleasurable dining experience.

Utilities, by definition, limit the choices that are provided to consumers. In the restaurant industry, it would not be viable for a business to offer customers every possible combination of ingredients so as to give complete freedom of choice in terms of menu. Consumers understand this and, as a result, expect a limited choice in terms of what can be ordered. It is important to recognize that, in areas of service delivery that have not traditionally been managed as utilities, this lack of choice can have serious cultural and political implications when a utility model is adopted. Emphasis should be given to ensuring that the requirements coming from the business are precise, a requirement of 'fast' is insufficient as it is then impossible to know when the requirement has been met.

A note to IT strategists

Traditionally, the IT department has usually articulated its value and differentiated itself from competition based on conversations concerning technical functionality. The utility model will change much of this, since utilities focus heavily on quality of service and much less on technical detail and function. 'Low-hanging fruit' in IT that suit a utility model are likely to include those where the consumer has stopped caring about the 'how' and is far more interested in the 'why' – followed closely by the 'when' and 'how much'?

Service level agreements

There has been much talk in recent years concerning the value of service level agreements (SLAs) within IT service delivery. Indeed, the importance of SLAs and their structure and value will be explored in detail later in this book. But what is an SLA and why does it exist? Let's go back to the restaurant...

On visiting the restaurant, patrons will bring with them certain levels of expectation with regard to quality of service. In the service (utility) business, this can bring untold dangers, since everybody's definition of 'good service' can and will differ. For this reason, it is a good idea to create a written description of the service that tightly defines what users of that service can expect. This written description is often referred to as a service level agreement (or SLA) and its use within IT utilities is absolutely fundamental to their success in terms of customer satisfaction.

In the restaurant business, rigorously defined SLAs are less common. This is largely due to the fact that most consumers have formed a pretty strong idea as to what 'poor' service feels like, and the maturity of this type of business is such that this view is largely shared by all patrons and restaurant owners. However, one might often find a written description of, for example, how the food tastes or what ingredients will be used to make the dish (you would not expect to order a hot lamb curry and receive a mild chicken one!). This would represent a crude and simple example of a service level agreement within our restaurant analogy. At various walk-in fast food restaurants, the business guarantees delivery of an order within five minutes (or you receive your order free of charge). Again, this is a crude example of SLAs (and performance penalties) at work in a restaurant.

A quick note on IT SLAs

In IT, after calling support, the SLA is the first place that consumers will go if they suspect trouble. Be careful how you define your service (and how aggressively) or it will come back and haunt you. More on this later...

Cost of service

Of course, the cost of a service is of great importance to consumers. Often, it is the driving force that compels a consumer to move to

Figure 4.3 Chargeback: what does it actually cost and therefore what price could you charge?

a utility in the first place. The cost of *providing* a service is also of vital importance to the utility provider since, without an accurate view of this information, it is impossible for a provider to ensure that they will not lose money once a service is established and provided.

Significantly increasing the cost of a service once it has been established can prove to be very difficult to justify to consumers. It is, therefore, important, during the service creation phase, to assess in some detail how much a utility service is going to cost to provide. Once a cost model has been established, it should be a reasonably simple task to decide how much to charge consumers for it (Figure 4.3).

In the restaurant business, the cost of providing a service is made up largely of the following elements:

- cost of premises;
- cost of staff;
- cost of stock (goods).

The premises and staff costs would normally stay relatively static (certainly once the restaurant reaches a 'steady state' and is established). The costs of buying in goods (particularly perishable ones) and utilizing these goods in a timely fashion (i.e. before they rot!) is largely what helps a restaurant to be successful financially. The 'supply chain' involved in keeping a restaurant appropriately stocked is of paramount importance and, in a good restaurant, is governed and maintained rigorously. Additionally, a restaurant may change the nature of the supply chain (maybe change a supplier in order to save cost, for example). In the utility world, this will not matter to the consumer as long as the service quality does not change.

IT costs in a utility model

There are valuable lessons for IT managers within the restaurant analogy. In providing IT service utilities we are, essentially, breaking down and reconstructing the 'supply chain' that manages the provision of 'raw' IT assets and turns them into something that users or applications can use. As we will see later in the book, a great deal of attention must be given to the total cost of ownership (TCO) associated with providing IT services in this way. Otherwise, there is a very real danger that, once established, the IT utility will not prove to be viable financially to its consumers, its providers, or both.

Service management and delivery workflows

We mentioned earlier that a large part of a restaurant's focus, in terms of cost containment concerns the management of its supply chain. Restaurants also contain their costs by ensuring that a service is managed and delivered using precisely defined, standardized techniques. Standardizing the way in which a service will be provided serves two main purposes. First, it ensures that a quality of service is maintained (if we deviate from a standard technique this will very likely change the quality of the service delivered). Secondly, the cost of providing a service is much more predictable if we do the things in the same way every time.

Of course, a restaurant closely governs the way in which it delivers its food by using recipes. These recipes are normally very precise and prescriptive and leave little room for the chef (or, importantly, the chef's staff) to deviate and experiment. A high quality restaurant will also employ very highly trained waiters and waitresses (who also work to strict processes and regulations) to ensure that food is delivered to the table in a timely and professional manner.

People who use utilities generally do not want surprises. Closely defining the way in which a utility service is managed and delivered is key to its success.

> **IT process governance**
>
> Typically, IT services today are not managed or delivered using standard processes ('workflows') and very little governance exists to ensure that standard 'best practice' is adhered to. Some industry standards exist to assist organizations in understanding how IT should be managed and delivered. Most notably the Information Technology Infrastructure Library (ITIL) is used by many organizations to apply discipline to IT service management. Standards such as ITIL can be used effectively to bring process standardization and governance to a utility computing transformation. ITIL and other relevant standards will be discussed in depth later.

4.3.2 Service deployment

The deployment of a service is key in terms of setting consumers' expectation with regard to service quality ('first impressions' and all that!). Restaurants have been known to fail almost immediately after their launch due to poor initial deployment of the service. In any utility, service deployment must be planned carefully in order to maximize initial consumer interest (Figure 4.4).

Of course, most of the focus and effort during this phase of utility enablement is on the procurement and installation of infrastructure. In our restaurant, this would mean moving into a new property, installing kitchens, customers' areas, toilets and so on. This is also the most costly phase of starting a new service, and so tasks and responsibilities must be planned very carefully (many restaurants have never even 'left the ground' and have gone out of business before opening day!).

One of the key challenges during this phase is to create an inventory that is appropriate to the first days or weeks of the service. In our restaurant, it may be difficult to judge just how popular the service is going to be in the initial weeks subsequent to launch day. From a cost perspective, this can prove to be a real headache. Often, one way of mitigating against the risk around deployment cost (particularly around inventory) is to negotiate flexible initial terms with suppliers. For example, a restaurant may ask a supplier to provide a certain amount of wine, but provide an SLA that ensures

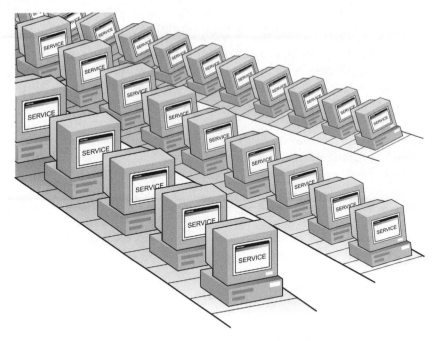

Figure 4.4 Rapid service deployment improves the chances of success.

that additional supplies can be provided within a certain timeframe. Remember, your suppliers are providing a service too!

IT infrastructure installation

Even after all these years, we are still not very good at IT infrastructure procurement and installation. Very few standards or 'best practices' exist allowing us to plan and install our new IT efficiently and without business interruption. In enabling IT utilities, much focus and attention must be given to the way in which the utility's infrastructure is sized and deployed initially. Otherwise, consumers, confidence will be low from the offset, leading to the utility service failing at the first hurdle.

The utility deployment phase is also the time at which definitions of *how* the utility should operate from day to day should be authored, learned and tested. Operational procedures, such as inventory and configuration management, service monitoring and incident management, should be considered carefully, written down and learned by the staff that will be responsible for carrying them out.

The deployment phase is a great time for rehearsal. Often, a newly established restaurant will invite close friends and family along to a 'pre-opening night' so that the service can be operated and managed using a friendly set of consumers to critique the service quality without fear of retribution if things go wrong. The importance of the friendly consumer cannot be overstated. Utility services will always be more successful during and directly after their launch if 'dress rehearsals' have been carried out, operational procedure fine-tuned and staff trained properly and immersed in service operation prior to launch day. Checks should be made that published service levels and provisioning (cooking) times can be met, and that there is always a fall back plan if the worst comes to the worst. In the case of a restaurant, sending out for fish and chips on the opening night would spell disaster; having to send out on the 'pre-opening' would be looked back on as a humorous anecdote if the opening was subsequently a success.

A friendly consumer for IT

In our experience, IT utilities should be launched using a 'pilot' project. This involves testing the service with real users who have been instructed to use and critique the utility without negatively influencing the consumers who will ultimately use the service. A pilot project is a great way of creating positive initial perception of a new service, but 'testers' must be managed carefully!

4.3.3 Service provisioning

So, the service has been defined, SLAs exist and can be measured, the infrastructure that will support the service has been installed and tested, operational procedures have been written down and staff have taken on their various roles to support the smooth operation of the service. Time to allow consumers to request delivery of the service so that it can be provisioned to them . . .

In a restaurant, provisioning of the service (i.e. seating customers, showing the service proposition (menu), taking their order, serving their food and billing them) is a complex and multifaceted workflow. As mentioned earlier, the only way of ensuring that the workflow is designed appropriately and that the service can be operated well under all conditions, is to rehearse during the service deployment phase.

To watch a busy, well-run restaurant in action is a sight to behold! A complex workflow is executed by highly professional, well-trained staff and everything is given direction and focus by the Head Chef (looking after the preparation of food for consumers) and the Head Waiter (looking after delivery of the restaurant's food).

Service management workflow in IT

Using the example of the chef's and the waiter's responsibilities in the restaurant, it is interesting to link this to organizational responsibility in the context of IT service delivery. In the IT service management workflow, we often break down the elements into 'service management' and 'service delivery'. Service management being the preparation and management of the service ('the kitchen') and service delivery being the actual provisioning of service to the client (the waiter delivering food to customers). The ITIL standards define well the specifics of each of these two elements. This will be explored fully later.

A service provisioning workflow in any utility will very likely contain some common elements, whether the workflow is supporting a restaurant or an IT utility. Automation of this workflow will reduce the associated costs (see Figure 4.5). These common elements typically exist as:

- request for service;
- approval to provision;
- resource allocation;
- service level allocation;
- cost model allocation;
- service provision notification;
- enablement of service level monitoring.

All of these elements will be discussed in detail later in the book. In both the context of our restaurant service and an IT utility service, the 'resource allocation' piece, in particular, can be very challenging and difficult to get right. Utility resources should be very well utilized, easy to reallocate and very quick to provision. If this is not the case, then this will have a huge impact on the efficiency and effectiveness of the utility model.

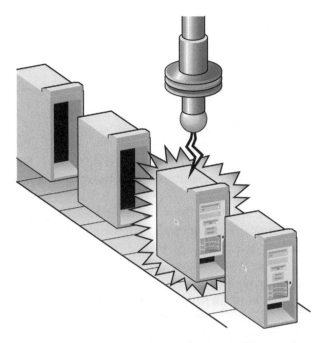

Figure 4.5   Automation is key to reducing the cost of the service.

Automated provisioning in IT

The challenges surrounding the provisioning of IT assets (such as servers) in moving towards a utility model are well known in the industry today. Many vendors can provide software solutions that can help in this area. Our experience is that these tools are of little, or no, value if they are not used with the rest of the provisioning workflow model in mind.

4.3.4   Service assurance

Service assurance activity ensures that, once a utility is actively providing service, its quality remains acceptable to its consumers (normally within the context of any service level agreement that exists). Some common elements that we would expect to find making up a service assurance workflow would include:

- usage monitoring;
- service delivery management;

- performance management;
- availability management;
- capacity management;
- problem management;
- service support;
- consumer administration.

In writing this book, we have formulated this list of elements with IT service delivery in mind (and each of these will be discussed in detail later in the book).

In a restaurant, much of the work that ensures good customer service is, once again, handled by kitchen and waiting staff. Kitchen activities such as stock control (capacity management and availability management) are 'background' activities that the customer does not necessarily see, but certainly feels in terms of quality of service (if a dish is not available, for example). 'Customer-facing' activities, such as checking on customer satisfaction (service delivery management) and presenting the customer with a bill (consumer administration), are things that the customer will experience personally.

Whose service are you assuring?

When creating IT utilities, we should be very clear about who the customer is in order to tackle service assurance in the right way. For example, great system backup performance will matter a great deal to an IT manager, but not at all to the HR application user. Be careful not to 'over-deliver' functionality in this area, but get to know your customer and focus on the things that *really* matter to them.

In some instances you may want to look at strategies akin to dual-sourcing in order to meet required SLAs, especially if you are relying on single suppliers with which you do not have an SLA yourself.

Interestingly, some elements of service assurance in a restaurant would actually span both the kitchen and waiting business units. Performance management, for example, cannot be handled effectively if both sets of staff do not work together in a coordinated fashion. A dish could be delivered very quickly by the kitchen staff, but if the waiter is slow to collect and deliver the order to the customers' table, then the restaurant will have failed to deliver good performance (and an SLA could be breached, leading to poor customer satisfaction).

Whether 'customer-facing' or not, all elements of the service assurance workflow will have a positive or negative effect on customer satisfaction, and ultimately sales, if not managed correctly.

4.3.5 Service reporting and billing

As we mentioned earlier in this chapter, for a utility service to be a success from a financial perspective it has to run efficiently (in order that its assets are well utilized) and it must keep a very close eye on the quality of service that it delivers (so that customers continue to use the service). Ultimately, both of these things lead to improved profits for the business, either because prices can be offered that are competitive (so attracting more customers) or because premium rates can be demanded (because quality of service is higher than elsewhere in the market).

Well-designed service reporting is critical since, through use of reports, we can gain insight into how well (efficiently and effectively) our utility is running. Reporting also allows us to make adjustments based on our findings (price changes, stock ordering decisions and so on). Equally, good service reporting procedure can be very effective in influencing customer behavior. In our restaurant, for example, we may be able to use our reports to show customers how successful our restaurant has been in the past (by showing growing customer numbers or reporting on customer feedback). In addition, the customer's bill is a report. If the bill is not generated quickly and accurately, then this can lead to a negative effect on customer satisfaction.

Some common elements of service reporting within a utility are:

- usage reporting;
- service level reporting (performance and availability);
- service billing;
- service accounting (profit and loss);
- utilization and capacity reporting.

Ideally, a business will utilize one centralized reporting facility that includes all of the elements listed here. Huge inefficiencies have been known to creep into organizations that have not centralized and standardized their reporting mechanisms. Incomprehensible, disjointed or inaccurate reporting can be more dangerous than no reporting at all. Reports should be designed that show clearly only that data that is necessary and that can be manipulated easily without the need for masses of re-engineering. Businesses change, customers change and marketplaces change. Reports must be dynamic and agile to support a dynamic and agile business model.

Some experience so far...

Our experience in designing and building IT utilities to date indicates that most internal IT organizations wish to focus less on service accounting and billing than on service level reporting and capacity management. This is because the organization wants to be seen as a 'value center', rather than a 'profit center'. It is our belief that this trend will change during the next 5–7 years and that, in the future, many IT departments will essentially be run as profitable (or at least full cost recovery) businesses in their own right. Of course, organizations that are implementing the IT utility model today in order to provide external consumers with service, must focus heavily on service accounting and billing.

Even the IT organization that does not want to 'turn a profit', will need to work hard to gain a detailed insight into service costs as, without this information, it may be hard to justify next year's IT budget. Interestingly, once set up, the utility-based ITO will not necessarily need its own budget, as it will, in effect, be set by the lines of business and the cost of services provided will be covered fully.

4.3.6 Service redefinition and retirement

A successful, established utility will not remain that way for a prolonged period if it does not respond to change. All too often, service businesses are seen to 'rest on their laurels' subsequent to periods of sustained growth and success. This leads to a move away from the very methods and disciplines that made the utility a success in the first place.

The world changes, market dynamics change and, most importantly, customers' demands needs and preferences change. It is, therefore, vitally important that a service-oriented business constantly measures conditions and adjusts its business model accordingly. Part of the benefit of the utility model is that 'customer-facing' attributes (price, service definition and so on) can be modified much more easily than in more rigid business models.

Our restaurant, for example, may decide that it needs to 'redifferentiate' itself after two years of successful operation. Maybe its success has led to new competitors introducing themselves to the market, looking for a piece of the action. Under these circumstances, the restaurant may utilize some of the market research and customer liaison techniques (discussed in Section 4.3.1) to re-evaluate new ways that a differentiated service could be created and deployed. Maybe, in the case of our restaurant, this means a completely revised menu or a refurbishment of the restaurant building itself.

This constant attention to customer preference and adjustment to value proposition is critical to a utility's prolonged success.

Using a utility model, it should also be possible to retire services completely and then re-utilize the assets that supported them in order to create something new. Our restaurant could change fundamentally from being an Indian restaurant to being an Italian restaurant. It should still be able to use many of the same methods, disciplines and assets. Good waiters are good waiters, a good kitchen is a good kitchen and good chicken is good chicken!

4.4 SHOULD INFORMATION SERVICES BE ANY DIFFERENT?

This chapter has attempted to introduce the reader to the world of the utility service, using the analogy of a restaurant to show how the various methods, best practices and processes that form a utility workflow can be applied.

The way in which information services are delivered to a business, we believe, has to change fundamentally during the next few years in order for businesses to continue to invest in information technology at anywhere near the levels experienced to date. It seems to us that a full implementation of a utility model (encompassing all of the areas discussed within this chapter) is *not* right for every area of information service delivery today.

Utilities are about quality of service, not functional uniqueness or leading-edge technology.

It is our belief that a full implementation of a utility model will bring significant advantages (around quality of service and cost control) to businesses when applied to the areas of IT that are well suited to the model. These areas will exist largely where the users of the service are more interested in quality of service than the specifics of technology enablement. In our experience so far, some 'low-hanging fruit' that exists today within the data center includes:

- enterprise data protection;
- disaster recovery enablement;
- enterprise storage;
- server management.

It should be recognized, though, that many of the methods adopted to create a utility model (good service management, abstraction of complexity through virtualization, tight cost control and accounting, for example) can bring significant benefit to *all* areas of today's data center.

4.5   CHAPTER SUMMARY

- Utilities live or die by providing truly differentiated services to well-defined marketplaces.
- Operational efficiency will have a large influence on how successful a utility becomes.
- Think about your differentiated service definition first, then process and then technology.
- Utilities will work for IT only where the consumer has stopped caring 'how' and just cares about 'why'.

5

Service Level Agreements

5.1 OVERVIEW

Service level agreements (SLAs) are critical to any utility computing project and while the concepts have been around for a while, they are only just beginning to find their way into the IT organization (ITO). Designing a good service and putting a suitable SLA around it is crucial to the success of the business as well as the ITO. Delivering IT as an effective service also means a change in relationship with suppliers. This chapter is designed to highlight good practice in creating services that can be delivered as a utility.

5.2 AN EVERYDAY EVENT

Service level agreements (SLAs) provide a cornerstone of any utility computing project. There is one cardinal rule for creating SLAs:

Make it easy to understand.

This manifests itself in a number of ways:

- a description of high-level customer deliverables;
- use of appropriate terminology;
- ensuring that all the key objectives of the SLA are measurable;

Delivering Utility Computing.   Guy Bunker and Darren Thomson

Our Gurantee...
Archie's
Pizza
Shack

Tasty, hot and on time...
OR IT's FREE!!!!

Figure 5.1 Service level agreement for a pizza chain.

- making the measurements objective rather than subjective;
- removing as much complexity (detail) as possible.

Before looking at how SLAs can be defined for IT and how to set about creating them, it is worth examining two that are probably familiar. The first is one created by a fast food pizza chain for its delivery service (see Figure 5.1). Often, this 'agreement' becomes a slogan or a company mantra, with it being printed on the side of boxes as well as being shown prominently in the shops.

It is very easy to understand, there is minimal detail, but it does appear to break the rule of objective, rather than subjective, measurements. What does 'on time' mean, and surely 'hot' is very subjective? For the former, there is some small print posted in the shops, or told to the customer when ordering a take away, which says within 30 minutes. As for 'hot', this relies on there being a well-accepted set of parameters for what it means; saying 'between 30–50 degrees Celsius' would cause more confusion than good, and so 'hot' is acceptable. Internally the term 'hot' can be made objective, measurements can be carried out to ensure that it is within the appropriate parameters, but it is not necessary to expose the customer to these details. Making it qualitative means that it is understood, which is the most important attribute, or service level objective (SLO) of an SLA.

One other piece that this SLA brings in is the idea of a penalty for non-compliance. While pizza chains are willing to make the guarantee and back it up with a penalty, the ITO will not be – at least not at this time. The penalty clause of the pizza chain and the parameters that they measure have been defined and refined over a large period of time. Technology has been created and deployed to ensure the penalty is not invoked very often. Sufficient scooters for deliveries at peak times are employed, hot boxes and insulated bags ensure that 'hot' really means hot, criteria for a delivery range (i.e. if you are twenty miles away they will not stand by their guarantee) are all used. While the ITO may not be willing to sign up for SLAs with penalties, it will only be a matter of time and confidence in the system before they are introduced.

Another example are the contract details found when looking at a network/internet service provider (ISP). Typically, other than price, there is a lot of information about performance. What will the average latency be between switches or routers? What will be the packet delivery rate and connection availability? For many users, knowing that the average latency will not exceed 50 ms is not particularly useful – unless compared to another provider. Then, 30 ms would be better than 50 ms – but is it useful? It would be if the application was streaming video, but not necessarily if it is an HR application. As with the pizza chain, the agreement contains penalty clauses if they are outside any of their parameters. For example, they credit the customer's account if latency on transatlantic networks exceed 100 ms over the course of a month.

However, the ISP also brings in additional parameters surrounding customer care and details on when to expect requested new services to become functional. This gives the supplier the opportunity to set expectation correctly and give themselves flexibility for items such as maintenance. Unlike the pizza chain, the service delivered by the ISP is continuous and so there is a certain amount of 'smoothing' that goes on; this ensures that every spike in performance does not raise an exception or SLA breach. When designing SLAs it is good to take this into account and give SLOs upper and lower bounds where appropriate.

From the perspective of the ITO, if it is going to be successful at delivering IT as a service then it needs to learn from everyday SLAs.

5.3   DEFINING SERVICES AND THE SERVICE LEVEL AGREEMENT

Defining the first service and the SLA by which the ITO is going to operate is daunting, but understanding what makes a good service and developing a fair but enforceable SLA will improve the chances of success by reducing ambiguity. Putting a process in place will reduce the risk of getting it wrong and provide the mechanism for rolling out new services in the future.

Before even starting to define a service, the first step is to talk to the customers and other key stakeholders. It is important to ask questions in order to really understand their needs. Successful services will be those needed by the customers; forcing badly designed services on them will result in failure for the ITO. Turning the ITO into a service delivery organization will require personnel with a new set of skills (see Sections 11.2.3 and 12.3.3), ones which can map business needs to IT and translate between the two very different sets of terminology. Discussing the needs with the customer will enable the ITO to build up a picture of what the business needs and develop potential solutions. In essence, a service is made up from using resources that have been configured such that specific features operate at specified levels (see Figure 5.2).

Using this as a starting point, it is immediately possible to see that services should be defined using existing, or planned for, resources. There is both a top-down, customer driven, and a bottom-up, ITO-driven process that needs to happen.

Leaving service design solely to the consumer will result in unobtainable goals. Leaving it to the ITO will result in services that no-one wants, but at least they will be built on realistic assets. The design of the services needs to be driven by the ITO saying 'this is what we can provide', while listening to the business requirements and then iterating through the service design until something agreeable to both sides has been decided. Within the service there will be smaller components, service level objectives.

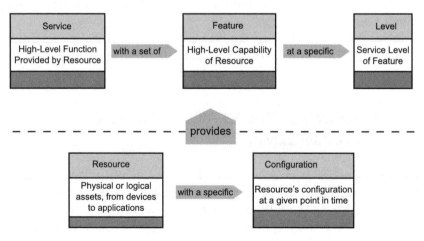

Figure 5.2 Relationship of service to resources.

5.3.1 Service level objectives

Service level objectives (SLOs) are the constituent parts from which an SLA is created, and at the lowest level they are measurable. If they go outside their parameter range, then this would cause the SLA to be breached. It is important to note that SLA penalties would probably not be invoked just because an SLO has been breached. For example, referring back to the ISP, because the latency between two switches or routers falls outside the original specification of 50 ms, this will cause an event to occur; however, it is the average over a period of a month that is actually important. From the consumer's and supplier's perspective, it is interesting to know that the service was running more slowly than expected, but this is for a point in time, not an average.

As long as the lowest level of SLOs can be measured, an SLA can use them and abstract them at the same time, so that the ITO can use the consumer's terminology. Storage performance can be measured in i/o's per second, but can be abstracted to 'fast', 'medium' or 'slow', or 'gold', 'silver' and 'bronze' depending on what makes sense.

While some SLOs are abstracted to make it easier for the customer to understand, many others are likely to be left unexposed to the customer. Often, they can be used as early indicators for potential problems, such as monitoring network speeds when looking for overall application response times. However, measuring everything that is available is also not advisable; there is a thin line between having enough information to be useful and too much as to strangle the resources available. If a metric is available but does not have any real impact on the consumer, or is of little use in resolving an issue, then it should not be measured.

While the goal is to have all pieces of the service capable of being accurately measured, there will probably be other SLOs that need to be defined. As with the ISP example, these will tend to be around the actual service that the ITO provides, for example, response time, not for the application, but for the ITO – how quickly will they respond to a request for a new server, or for a request to increase the amount of storage on a volume? By putting all of these items in the agreement, the correct expectation is set. It is essential to avoid terms that can be misunderstood, a response time by the support organization of 'good' is difficult to measure and one person might consider an hour to be a 'good' response, while another will become frustrated at waiting a few minutes. Setting an objective of say, ten minutes, will set the right expectation on all sides.

At the end of the process there will probably be a number of options for the ITO to realize the service, and the process may have led to a gradation of services that can be offered. The differentiator for them will be cost.

5.3.2   Pricing the service

As discussed previously, utility computing is all about money. The pricing of a service is very important, not only for the ITO to know how much it will cost to deploy and manage, but also for the consumer to help them make a good commercial decision. When looking at a menu in a restaurant price is often the first item that is viewed. The more expensive the item, the 'better' it is, although it might not suit the palate of the customer, or the wallet.

When pricing a service, capital expenditure is only a small part of the equation. Maintenance and reinvestment costs, in addition to general management, also need to be included. For example, if running backup as a service, the backup can be automated fully and all the appropriate costs apportioned, but what about restore? How often does a restore occur, how time consuming and manually intensive is it? These hidden costs are frequently forgotten. While it is not an aim to turn the ITO into a profit center, it should be an aim to measure accurately and apportion all the costs associated with a service. In some cases, the SLA may have requirements put on the customer, for example, that no more than one restore can be requested at any one time. It might be that if multiple requests are made, this could cause a degradation in the service provided.

Note for service designers

A complete SLA will have pricing in it, even if it is just a rough approximation; without it, the users will always pick the best option (i.e. the most expensive). Similarly, a service and an SLA cannot just be a price, at a minimum there needs to be a description of the service and how it differs from other services through functionality, or the levels of functionality offered.

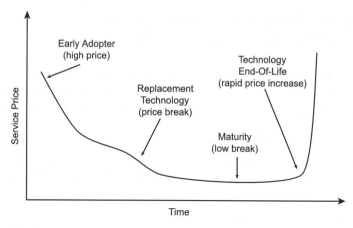

Figure 5.3   Service price over time.

The price of a service will vary over time (see Figure 5.3) and should initially decrease as more consumers take up the service. Take a storage service as an example. Over the course of a year, the hardware associated with the service will reduce in price or, more likely, the cost per GB will decrease as disk capacity increases. New technology will decrease the price of older versions and, for consumers, it would seem like those using the new hardware would get a better deal on price. It is up to the ITO to average out prices to the consumers (while maintaining accurate costs). There is no good time to introduce pricing changes, whether these are up or down, but having a fixed changeover date and adequate warning helps lines of business plan for the forthcoming year. Just as the cost initially comes down, so eventually it will go up. This could be for a number of reasons: manufacturers removing support for specific pieces of hardware or software versions, 'old age' of hardware, which has reached its lifetime limit and needs to be retired. At this time, the management of the service will start pushing the price up; migration to a newer service offering will probably be preferable to the consumer, rather than the increasing price. One way to mitigate the cost of retiring a service and moving users to a new service is to hold back some of the price paid (over-recover) to enable the reinvestment to take place. Care should be taken as this can cause issues; the ITO may appear to be making money from its users in the short term, or worse, the money 'saved' may be removed from the budget in the future as it is seen as being unspent. Upfront explanation of all costs is essential to build trust in this new service environment.

5.3.3 The miscellaneous pieces in a service level agreement

While the bulk of an SLA is made up of the service being delivered, there are a number of other pieces of information which are useful to gather and put in place. Going back to the ISP example and building on it, these could be:

- name and contact details of the customer, including telephone number, email address, location, group or line of business;
- secondary contact details;
- date from which service is to start;
- date on which service is to finish (if applicable);
- pricing/cost details – even if there is not going to be chargeback;
- support objectives, how quickly will the ITO respond?

  - is it 7 x 24 support, 9 x 5, or email only?
  - what if there are changes in requirements, for example more storage?
  - support contact details; is there a telephone number? What about out of hours? Is it a person's name, or a specific group?
  - is there a different response to different scenarios? For example, if the application is down, will the response be different from it running more slowly than normal?
  - are there planned maintenance outages, for example one hour per month on the second Saturday, between 5 pm and 6 pm?
  - a technology refresh commitment, how often will this happen, who will be responsible for migrating applications and data?

There are other pieces of information that should also be captured at this time:

- key business users, for example, finance department and shipping;
- business importance of the service, for example, business critical;
- additional areas for support, for example, network, database and server teams;
- third party suppliers for both hardware and software.

While this seems like a lot of information to capture just to create an SLA, having it will mean that risk analysis reports can be created more easily. Being able to report, and subsequently review, all services that are business critical will help, not only with service update planning, but also other areas such as disaster recovery. It can also be used to find other instances of services, which are not marked as business critical, but actually are.

One of the most common failings when mapping existing applications is that, while one server with the application is marked as business critical, another (on which it depends for data) is not marked as being critical. In the event of a failure, the critical servers are brought back online first, along with the applications, but the applications fail due to the dependency that was not spotted during the mapping.

5.3.4 Living up to expectations

Introducing SLAs to both the ITO and the consumer and then managing them is not an easy task. As with every paradigm shift, there needs to be proof that the new environment is better than the old one and the only way to do this is with reporting. In addition, there needs to be proactive management to catch the problems before they become a real issue.

Instrumentation in both hardware and software has improved in the last few years. Server uptime is not adequate, as it is no proof of the application running, let alone whether the performance is acceptable. The only way to do this is to look at the application from the user's perspective and then trace performance back through the system to the assets being used, including the network links, and that might mean tracking all the way back to the physical disk the data is on (see Figure 5.4).

Technology exists to instrument applications[1] automatically to enable this type of data collection. Many services will not require this level of sophisticated monitoring but the principle remains – the customer is king and saying 'it seems to be OK for me' will not be sufficient for the new service-driven ITO.

---

[1] For example, VERITAS i3™.

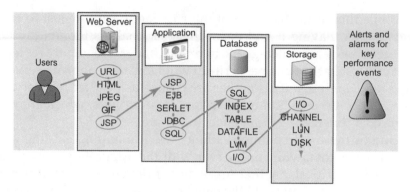

Figure 5.4 Performance analysis starts from the user's perspective.

> ### Note for service designers
>
> While all aspects – server, storage, network, application – are involved in delivering a service to the end user, a clear definition of who owns what is required. Problem resolution in a timely manner will rely on the complete team working together, rather than each pointing the finger at the other.
>
> The SLA need not reflect the interaction between various groups, but the process of doing so should be well defined and known to all involved.
>
> Where a service relies upon an external, or third party, service over which the ITO or organization has no control, this should be put into the SLA – again defining where the boundaries are and the potential consequences. Where possible, an SLA should be set up with the external service provider before it is used.

Measurement of performance tends not to be binary. More often than not, while the SLA points to maximum or minimum response times, from the perspective of the ITO, it is the trend that is important in order to anticipate potential service level breaches. By continuous monitoring of the service and storing the data in a performance data warehouse, degradation can be detected. Of course, detecting that the performance is deteriorating is only the first of step in the process of resolving the issue:

1. Detect application performance degradation.

2. Find the source of the problem.

3. Focus on the root cause.

4. Improve the response time.

5. Verify the result.

While the end-to-end approach from the customer's perspective is helpful in finding where the problem may be, it is only the start. Other, more traditional, tools for system and network monitoring can now be brought in to see if there are problems outside the control of the application which are manifesting themselves as potential service issues. One common example is that the Ethernet is being used to capacity by a backup or restore task, which in turn, is causing the application to run more slowly. Another example might be that another application on the server has taken all the CPU cycles.

Often, the root cause of the problem is different from the obvious one. Identifying the culprit is a skilled task, which, while helped by tools, still needs a proficient administrator to make the decision as to how to fix it. Once a change has been made, it should show up in the monitoring as an improvement, if not, then it was not necessarily the appropriate reaction to the problem and other alternatives may be needed.

For some SLAs, actions to potential issues can be automated. For example, if a file system is supposed to remain between 75–85% utilized, a policy can be set to grow or shrink the file system appropriately. This brings to the fore another issue, 'spike' management. With the file system as an example, the policy should take into account any potential spikes that might occur. If the value were to just hit 85% before falling back, then automatic provisioning should not occur. The same should be true of performance monitoring, there need to be additional parameters set by the ITO to ensure that additional unnecessary work is not created.

Note for service designers

While this example is designed to illustrate growing and shrinking a file system, with the storage coming from a pool that is owned by IT, it could, in a fully realized utility ITO, come from a supplier-owned pool. In this case, the contract with the supplier needs to support decreasing, as well as increasing, usage. This is true of applications as well, where licenses may well need to be deployed, redeployed or removed 'instantly'. The onus should be on the supplier to meter usage of their products, rather than insisting the ITO does it for them.

Effective service management will not only help the ITO live up to the expectations set, it will also result in happy customers.

5.3.5 Chargeback

Introducing SLAs to both the ITO and the consumer, and then managing them, is a big step and a big commitment on both sides. However, there are other reasons for doing this. At the end of the day, it all comes down to money and it is essential for both the ITO and its customers to understand the costs associated with the services provided.

While visibility into costs and where the value to the business lies is important, it is also very easy to fall into the trap of attempting to measure cost down to the smallest detail. This results in people becoming obsessed with apportioning cost to everything, from bytes across the network to unused CPU seconds on servers. Unfortunately, there is no simple answer; however, keeping service costing as simple as possible is a good mantra to maintain.

There are three basic costing models:

- *Fixed*. For example, use of a server, or per user, or per transaction.
- *Fixed and variable*. For example, a fixed fee for a backup service with a variable component based on how much data is backed up or restored.
- *Variable*. For example, online storage usage.

The backup and storage utilities are now becoming commonplace and the chargeback formulas for them are well known. For online storage, many organizations carry out a 'true-up' at the end of the year, taking into account how much storage was bought, and factoring in how much legacy storage exists and how many administrators administer it. Simple arithmetic enables the fairly accurate pricing of delivering storage as a service over the next year.

Servers are more problematic and applications even more so. When a server is 'owned' by a single line of business for use with a specific application, it is easy to apportion the cost. However, if that server is used by multiple departments, running multiple applications, then the problem of apportioning costs escalates. CPU usage is the most common way to charge back server usage; the delivered compute minute (DCM) or delivered software minute (DSM) being the most common metrics – the latter having finer granularity as it can be associated with specific applications. Amalgamation of CPU usage across multiple tiers is now commonplace when looking at total cost for an application. In order to overcome issues relating to the relative power of the CPU, an additional normalization multiplier is required. SPEC[2] publishes numbers which can be used for just this purpose; alternatively, an in-house application can be developed and used for this purpose.

Other metrics can also be used to create more detailed chargeback information, such as memory and swap space. However, the more metrics that are applied, the more complex the formula and the more expensive to manage and maintain it becomes. Keeping chargeback as simple as possible will pay dividends in the long run by not confusing the consumer.

In late 2004, Sun Microsystems Inc. announced that it would sell computing services based on CPU usage. The company has kept the pricing very simple to understand at $1 per CPU/hour. Of course what this has really done is made the ITO look very closely at the cost it incurs when delivering compute services, in order to compare it to this widely publicized figure.

The mainframe has always had CPU-based pricing and when examining chargeback, especially when servers become shared, it might be worth considering a variable rate, depending on time of day or day of week. The additional complexity may prove unnecessary, but when looking for further ways to drive out cost, and providing a flexible compute infrastructure is in place, peak-period pricing can help to level out usage. By making it more expensive to run applications during peak business hours, non-essential applications (such as reporting) can be pushed into low-load times. With the advent of virtual machines, this might also equate to reducing the resources available to the application during the day and then increasing them during the night.

Ultimately, perhaps, the ideal customer-friendly chargeback process would be related more directly to the application, perhaps relating the number of transactions to the resources used. Transactions could be as simple as web pages served out, or they could be database transactions; by tying cost down to useful quantities of work, the business will be able to calculate the real value of the IT services it is using.

There is a dilemma that the ITO needs to resolve. Under the new governance of being able to purchase and deploy hardware and software fairly independently of the various lines of business, there are obvious savings to be had from the control it gives, whether it is to flatten the purchase cycle, or just to buy in bulk. However, services need to be charged appropriately, even if only part of a server or disk

---

[2] Standard Performance Evaluation Corporation, http://www.spec.org/.

array is being used. Loading of service payments based on partial occupancy, while practical in the eyes of the ITO, is not a good way for a service business to conduct itself. That having been said, the costs due to bulk purchase should be lower for all from the outset. One way to ameliorate the problem is to partner with vendors who will share the risk and deliver their services based on need and usage (see later in this chapter).

Special services, such as those that do mean sole occupancy of hardware, should be priced accordingly. While it is not an aim to turn the ITO into a profit center, it should become responsible for losses due to poor accounting.

Many businesses that are not interested in actual chargeback, but still want to give the various lines of business insight, are now creating 'information only' invoices[3]. These give all the information relating to the cost of services provided, but no money changes hands. Being able to attribute usage and the business value provided from the service should not be underestimated and is good to have during talks on requirements for the subsequent year's budget.

Accompanying the chargeback report there should be a service report, detailing how the ITO has performed against its goals. For those ITOs not prepared to sign up for an 'agreement', a service level expectation (SLE) can be used instead. The goals remain the same; however, the monetary exposure for missing terms in an agreement would not be present. Over time, the SLE can be converted to an SLA when all parties are sure that the correct monitoring and reporting infrastructure is in place.

A cautionary tale

Back in the hazy days of the late eighties and early nineties, a large company decided to centralize its IT department. It would provide IT as a service for open systems (yes, it was years ahead of its time) and would charge back for usage – let's call the internal currency 'Orange Bucks'. All went well for the first few years and then departments started to put their own IT in place rather than using the central service. The reason was twofold: first and foremost it was cheaper to use real money, 'Green Bucks' to buy the equipment and services, and secondly, the department could respond more quickly to the changing business requirements.

---

[3] Also referred to as *shadow accounting*.

5.4   CREATING COMPLEX SERVICES

Services should build upon services. If care has been taken to design two or three storage services, and two or three server services, then higher level applications should build on these. If the higher level applications cannot use the lower-level ones, then they should be changed. The ITO needs to have one set of rules – not one set for itself and another for its consumers.

Each service defines its own set of SLOs which, from a monitoring perspective, can be aggregated to create all the touch points for the service to be provided. A simple web service (see Figure 5.5) will need some storage, some compute power and some network. In the case of storage, it might be delivered as a file system, which was built on a logical volume and, ultimately, on an intelligent array.

Several different versions of the web server service may be designed to accommodate various user communities. An internal intranet site would not have the same availability requirements as the external e-commerce site, or even the internal HR one. Abstracting out the features to create a service that is understood by all is important (Figure 5.6).

In this case, we might just concentrate on the response time, network throughput and compute rate. Other comparative services

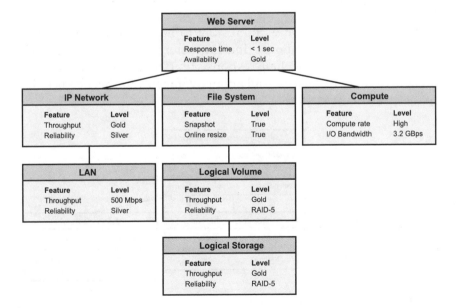

Figure 5.5 Building a web service.

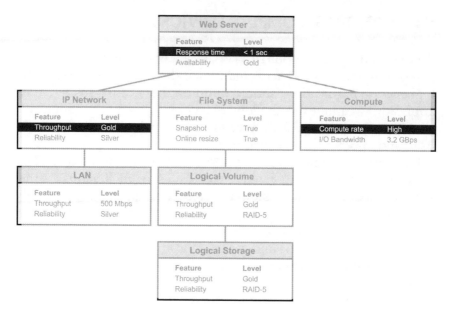

Figure 5.6 Abstracting SLOs to reduce complexity.

may have a slower response time, which would reduce the price of the service.

Once defined, this too could be used as part of an SLA for a more complex application such as a three tiered e-commerce solution (see Figure 5.7)

In this last instance, it is interesting to note that the sum of the constituent parts for response time is greater than the SLA. All three pieces could be within their response time SLOs and still cause a breach. In this instance, it is all about averages and this needs to be made clear to the customer.

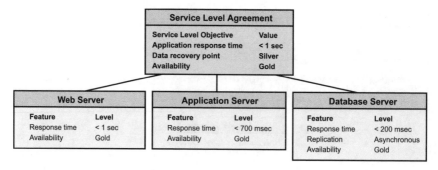

Figure 5.7 An application service.

> **Note for service designers**
>
> When designing an application service, metrics are tough to get exactly right. While all types of load testing can be carried out in non-production environments, it will only be when an application goes into production and the real loads are experienced that the true metrics will be found. For this reason, it is worth investing in tools that can monitor things like performance, so that any degradation in service can be spotted. When initially setting up the service, it will be worth having a period of time where values can be measured and adjustments made before the SLA comes into force. When the environment is changed, for example the server or application is patched or a new version installed, the level setting of the metrics will have to be done again. Discussing this upfront, will set the appropriate expectations with the consumer.

5.5   MANAGING SERVICES

Like all utilities, IT services have to be managed, and also like utilities, the services have a lifecycle (see Figure 5.8) with two distinct halves: service delivery and service management. Within these two sections there are three primary roles: the service designer, the system administrator and the operator. While these can be combined and, in smaller businesses, may well all be the same person, it is useful to separate them out.

Service definition is carried out by a design role, with the usual caveats of ensuring that the service is something that the consumers will want. The system administrators are also involved in this phase to make certain that the services can be delivered with the required attributes and within the appropriate time scales. Once defined, these are often placed in a service catalog, which is then made available to the consumers through a portal. It is important that services receive visibility within the organization and, as with most web portals, making it simple to navigate is important. By offering service requests through the web, the ITO can begin to build a community of users that help themselves. Clear, concise information on the service is required, along with price information, to help the user

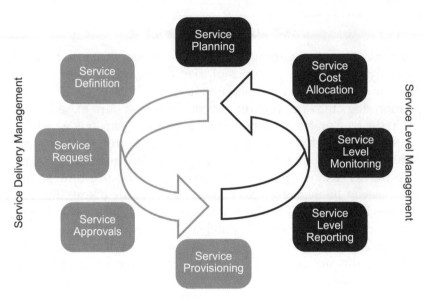

Figure 5.8 Service lifecycle.

choose which service is fit for their needs. Once they choose the service, there may well be additional information required, such as how much disk space is needed or the name of the web server. The user then submits the request.

Once submitted, workflow is used to automate the delivery of the process. In order to maintain control, an approval process should still be used. The actual provisioning of the service could be automated fully using workflow, or could have individual steps that require administrator intervention, for example checking that the storage or server selected is available. Ultimately, service provisioning should be carried out by operators rather than administrators, with all the best practice captured in an automated workflow. In the future, this may go one step further, with complete automation. However, in the near term, building trust in the automation tools used is important and so the operator or administrator needs to be involved to ensure that the request is completed properly. At the time of provisioning, monitoring and instrumentation needs to be deployed in conjunction with policy. The information collected forms the backbone of reporting and cost allocation. The consumer, who originally requested the service through a portal, can now return to it to monitor the current state of their services.

Reporting is critical for the ITO and for the consumer; as well as monitoring the health of the deployed services, it also provides a valuable feedback mechanism to assist in future service planning and a natural selection mechanism for removing those that are unsuccessful. Are there services in the catalog which are not used? If so, then why? How many (and the answer should be zero) services had to be modified before they were provisioned?

Reporting the usage of a service will also help in planning for subsequent deliveries of the service. Is there sufficient hardware and software to deliver the service? If not, how long will it take to get it should there be a request made? Trending is not always sufficient and additional information should be sought. Are there application or business changes that might affect subsequent service requests? Will the customers change their behavior because of the business changes?

For a business successfully to maintain a competitive IT edge, the ITO needs to innovate. Innovation is not about bringing in the latest and greatest technology, but understanding how the technology can bring value to the business and then wrapping a service around it so it can be deployed rapidly. Often, new technology is difficult to exploit as very few people within the ITO understand the impact it might have on the businesses and how it might be provisioned. Continuous discussion with the consumers of IT about their businesses and requirements can make sure that when new technology comes along, it can be utilized if appropriate. By creating template solutions, which can be deployed by anyone in the ITO, the technical advantage can be made available for all.

When designing and deploying services, consideration also has to be given to what happens to the service at the end of its lifecycle. A migration plan needs to have been created to move existing consumers onto either a brand new service, or onto a new version of the service. This will probably include migration of data as well as migration of applications. For complex services, this migration process could become complex and if it is something that needs to be repeated often, for example if there are several hundred database applications, then the creation of workflow to ensure best practice is followed during the migration should be considered. Part of the cultural change that needs to happen within the ITO is to look at any task that has to happen more than once, and then calculate whether investing the time to automate it is worthwhile – no matter how small the task.

The migration of data and applications need not be restricted to the end of life of an application. The business value of data and applications varies over time, by enabling the migration between different service levels, further services can be created and operated. This will enable customers to deploy applications in their early stages at one service level and then migrate them either up or down the service levels to ensure that the appropriate cost of the service provided matches the business value gained.

Technology refreshes are another cause for migration, applications can be running through several technology refreshes. Mainframe applications are a good example of applications that have lasted 10 or 20 years longer than expected. Often the migration to new technology is not an option, not just because the application will not run there, but because the skills and understanding of how the application was created and what it actually does have been lost. It is therefore easier to leave it alone, following the old adage of 'if it ain't broke, don't fix it'. Even when migrating applications from one server to another in just a hardware and/or OS upgrade, there needs to be extensive testing that the application performs correctly. Migration of the application to a different OS entirely (even if it is supposedly portable, for example a Java application) will need extensive testing. The application and its business criticality will define just how much testing is needed. In some applications, code might only be run once a year, for example at year end, so code coverage tools may need to be employed. Specifying upfront who is going to handle the testing in this scenario is best done in the SLA to alleviate potential finger-pointing problems later.

Ultimately the ITO needs to provide a customer relationship management (CRM) system (Figure 5.9). CRM has, up until now, always been applied to external customers, providing a strategy for understanding the customers and developing the strong relationships which are needed to maintain the business. From an ITO perspective, there is no difference – they need to build the same relationships, even though the customer is inside the same organization; the support organization will morph into a customer service organization and will include disciplines for marketing services, gathering feedback and watching for trends.

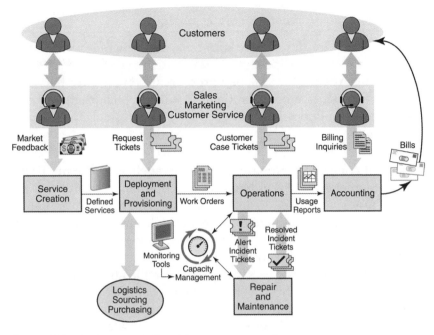

Figure 5.9 Customer relationship management.

5.6 SHARING RISK WITH SUPPLIERS

Embarking on a utility computing project will not just touch the ITO and the business; it will also need to include IT suppliers. Creating a responsive and agile infrastructure that can respond rapidly to business needs, but which is also cost effective, has its issues. For the ITO to be able to respond to requests for new services for storage and servers rapidly, it implies that there are spare resources to hand. However, having spare resources is neither cost-effective nor efficient, what sort of resource will be needed?

In order to have the resources to hand, service level agreements are needed with suppliers, without these, turning the ITO into a service delivery organization will not succeed. SLAs with suppliers will enable the ITO to create SLAs with its own customers, based on delivery SLAs from its suppliers.

> **Note for service designers**
>
> When relying on third party suppliers, and where appropriate, it is often worth considering a dual-sourcing strategy. This will enable the organization to keep pressure up on the suppliers, as there is competition between them. It will also ensure that there is a second option when delivering services under an SLA to consumers.

When it comes to hardware, companies such as IBM, Sun, EMC, Fujitsu–Siemens and HDS have offered on-demand solutions for the last few years. They install a machine, often fully populated, which is enabled incrementally through a license key or through internal system accounting. In doing this, they are sharing the risk around, whether the business will utilize all, or none, of their solution. When it comes to software, some applications are capable of incremental usage, however, the majority are not. This is slowly changing as customers request that software is purchased on pay-as-you-go type licensing agreements and automated license usage reporting is implemented.

There are a number of implications when investigating this type of arrangement with suppliers. The first is whether the service being bought can be scaled down as well as up. Most vendors are willing to increase the quantity of resource being requested, but not to reduce it. There are a number of issues surrounding usage reduction, not least of all revenue recognition requirements for vendors.

One other important point about usage-based pricing is how the company accounts for it. Under the normal business model, assets are capitalized over a period of time and written off against a profit and loss account. At the start of a financial year, it is obvious as to how much is to be spent and so plans can be made accordingly. In a usage-based model this will not be the case, as usage will depend upon demand – if the demand is greater than expected, then the associated usage will be higher, along with the costs. The vendors will also charge a higher price for the more granular usage to help cover their costs of sharing the risk. When looking at vendors who offer usage-based pricing, look at whether the usage is allowed to go down and examine the tools they have for monitoring usage, as those will become your new best friends.

Warning! Warning! Warning!

The decision to move to a utility computing, pay-as-you-go model with suppliers is not something the ITO alone can decide. There are deep financial ramifications for the business and this would be a decision the CFO and board would make – not just the CIO!

5.7 SUMMARY

- Service level agreements tie the IT organization to the business.
- Metrics found in SLAs should be measurable, although abstraction will allow for more subjective values.
- Trust needs to be built up by the ITO with the business, and this can only be done by asking questions and really understanding how IT is used to drive value.
- Services need to have a price associated with them to help customers decide what level of IT investment they need for the applications they own.
- In order to be successful, the concept of SLAs needs to be taken to suppliers in order to share risk.
- Usage-based pricing also helps to share risk, but prepare for new accounting practices to be put in place across the organization.

Part Two

Transformational Modeling

'Destiny is not a matter of chance, it is a matter of choice; it is not a thing to be waited for, it is a thing to be achieved.'
William Jennings Bryan (1860–1925)

In order to build the next generation of IT organization, which focuses on delivering IT as a service, a transformation has to occur. This part describes how to tackle the problem by looking at the business driver and proposing a Utility Computing Reference Model. By decomposing the model into its constituent parts, a complete methodology is given to enable the ITO not only to become more responsive to business needs, but also to become more efficient in the process.

As we have discussed in previous chapters, changing an area of IT so that it operates as a utility involves a complex and multifaceted transformational initiative that involves much more than technology enablement and revision. Many layers of an IT organization will be impacted by this kind of change, and all likely changes must be very well understood by project stakeholders.

To this end, we have created several artifacts that can be utilized as models during the formation of a utility computing strategy. These artifacts form a complete methodology for utility computing transformation. They include the following:

- a Utility Computing Reference Model;

- a Utility Computing Maturity Model;
- a Utility Computing Transformational Approach.

Together, these artifacts can be used to create realistic, compelling and pragmatic IT strategies that allow a business to move towards the utility model at an appropriate focus and pace.

6

Project Justification and Focus

6.1 OVERVIEW

This chapter will explore and describe some of the lessons that we have learned concerning the matter of how to compel a business to embark on a journey towards utility computing. As has been stated elsewhere in this book, few successful utility computing projects become successful without serious commitment from business executives who have (or have been given!) a passion and enthusiasm for developing IT so that it operates as a service.

Gaining an understanding of true business drivers is a good first step towards proposing utility computing in a way that demonstrates real business value to the people who will have to fund and sponsor a project whilst taking some risk in order to make its adoption a reality.

Similarly, we have found that some areas of IT prove to be better targets for transformation than others. This chapter exposes some guidelines for finding and targeting 'low-hanging fruit'.

Delivering Utility Computing.   Guy Bunker and Darren Thomson
© 2006 VERITAS Software Corporation. All rights reserved.

6.2 BUSINESS DRIVERS AND PROJECT JUSTIFICATION

Before looking at the three methodology artifacts, let's explore the things that, in our experience, typically compel an organization to deploy utility computing. This is critical to strategic planning, since the change to an IT environment that utility computing demands is significant and needs strong justification to be supported adequately by business executives.

It is important to recognize that the adoption of a utility computing method will happen at different rates (and with different business goals) for every organization. Our experience in utility computing projects so far has taught us not to make assumptions about what the IT organization wants to accomplish with utility computing.

Broadly, organizations with which we have worked have tended to want to drive improvement into the following three areas:

- control over IT costs;
- IT quality of service (QOS);
- IT agility and flexibility.

Although these three value drivers are common across most utility computing initiatives, the balance of these three varies greatly from organization to organization (and often depends on vertical market type).

For example, for one banking customer with whom we have worked, the primary business driver for utility computing was cost reduction–'reduce IT expenditure by $50M per year, at any cost (except loss of service quality)'. In a media company, where we recently deployed a utility for disaster recovery, the primary business driver was quality of service–'the business must be able to depend and rely on IT service and the cost must be appropriate'.

It is often useful to spend time exploring with the stakeholders of a transformational project the business drivers that exist to cause an interest in utility computing. Our experience is that it is very easy to create complete and well-designed IT utilities that do not deliver value in terms of a stated business requirement. In an organization that desperately needs to reduce costs, a newly deployed utility that fails to do this, but delivers great service quality, will be seen as a failure (Figure 6.1).

Figure 6.1   Misunderstand the requirements and you will miss the goal.

It is true to say that one of the attractions of utility computing is that it allows IT organizations to improve the three main areas of benefit in parallel. However, it is still vitally important to give appropriate prioritization to them, based on the stated business drivers that exist.

6.3   HOW TO FIND WHERE TO START

The success or failure of a utility computing transformation will largely, we have found, depend on what you try to transform. As has been mentioned previously in this book, finding some 'low-hanging fruit' (a well-defined and scoped area of IT to transform with maximum benefit and minimum risk) and using this scope to create a utility computing 'pilot', is generally a good idea. So, what represents good 'low hanging fruit'?

In our experience, there are some general rules that can help to target an appropriate area of IT infrastructure for utility computing transformation. It is important to recognize that finding a good target is not just about what can be accomplished with technology. Utility transformation is about people, process and service, and taking these factors into consideration is critical in hunting out a transformational target.

- Rule 1 – Utilities are about quality of service *not* uniqueness of function

  In other words, do not target an area of IT where its users still genuinely care about *how* a service is provided. Utilities, by definition, restrict the choices that users can make in the context of a service that is provided to them. Targeting an area of IT that is still changing constantly due to varying user requirements or aggressive technology innovation is probably not a good idea. Instead, find an IT service where users have stopped caring about the '*how* ', and where they are more concerned with the '*why* '. Our experience has shown us that the utility computing model works best where users are happy with a somewhat restrictive 'class of service' model, as long as they are provided with a consistent, reliable and well-documented quality of service.

- Rule 2 – Do not try to transform too much

  In creating your first utility computing service, you are already being bold by using a new technique in IT to transform radically something that is probably not fundamentally broken as far as its users are concerned. Keep the initial scope of the transformation as small as possible in order to give yourself a chance to focus heavily on creating something that is very effective. If you can demonstrate, having created a relatively simple IT utility service, that the utility computing model can bring real benefits to a business, then it is likely that the business will gain a huge appetite to use the model further.

  By way of an example, we have worked with several clients who initially targeted their storage networks for utility computing transformation. In other words, they wanted all offline, nearline and online storage to be provided as a utility with full class of service models, automated workflows and automatic billing ('chargeback') in place. Although IT storage does represent a natural fit for utility computing in most cases, our advice to these clients was to start with system backups and restores. Backup represents a subset of what would eventually become a full range of storage utility services, but it is far easier to transform than live production online storage services, its transformation has less impact on users and it can be transformed with much less business risk than online storage. All of the clients that we worked with to create a 'backup utility' later extended the model to encompass other storage services and, of course, they had learned some valuable lessons along the way.

Figure 6.2 A successful service provider – a friend for all.

- Rule 3–Make sure that the business case is truly compelling
  As discussed earlier in the book, all businesses will have slightly
  different motivations for introducing utility computing. Whatever
  the motivations behind a utility computing project are, it is absolutely critical that the initial target for transformation is capable of
  demonstrating satisfactory achievement of the business objectives
  that exist. This leads to another important point . . . make sure that
  there are some business objectives, that they are written down
  and that all project stake-holders agree with them. Make no mistake, there are likely to be people, either internally within your
  company or externally who will want to see the utility computing
  transformation fail. Do not give them ammunition; create a compelling business case for the utility model, achieve the goals that
  this document laid out and then shout from the rooftops about
  your success! (See Figure 6.2).

- Rule 4 – Do not start until executive–level sponsorship has been
  attained
  Utility computing projects are not generally successful unless
  they are endorsed fully and supported by one or several senior executives in the business, and where these individuals are passion-
  ate about making utility computing successful. This is due largely
  to the cultural implications that the introduction of the utility
  model has on a business (the impact, for example, of user choices
  being taken away or lines of business being charged for services).
  Ensure that, once your transformational target has been decided,
  your executive sponsors are fully supportive, are in agreement
  with your choice, and believe the business case that is associated
  with the project.

7

The Utility Computing Reference Model

7.1 OVERVIEW

When transforming anything radically it is generally a good idea to know where you are going. It seemed to us that nobody in the IT industry had really created a complete (and unbiased) 'go to' model for utility computing.

In thinking about how to support organizations that wish to create utility computing strategies, we made a decision very early on to create and document a reference model. This model was created for four main reasons:

1. To dispel the speculation and ambiguity that currently surrounds the use of the term 'utility computing'.
2. To articulate clearly the philosophy and principle of the utility computing method.
3. To provide a context in which to evaluate and determine the value and impact of adopting utility computing.
4. To provide a 'go to' model with which to compare a current IT estate so that transformational plans can then be authored.

Importantly, the Utility Computing Reference Model has not been designed in a prescriptive way. We have found that, generally, our view of 'the way that an IT utility should look' tends to work well

Delivering Utility Computing.   Guy Bunker and Darren Thomson

for most organizations. However, many projects focus on using only some elements of the model and others change the model to suit their own particular requirements.

In essence, the thought that went into the creation of the model was born out of a belief that to focus solely on the 'mechanical' issues of IT (such as the hardware and software) dilutes the benefits that can accrue from utility computing. Indeed, recent project experience has demonstrated to us that the true value of utility computing may lie in the service orientation and organization transformation that it facilitates.

The Utility Computing Reference Model (See Figure 7.1) is designed to provide a practical framework for the introduction of utility computing and is based on experience gained in real-world projects, together with research and theoretical guidance from IT and business transformation specialists. The model is a 'how to' guide for organizations wishing to adopt a utility computing paradigm. It attempts to be holistic and deliberately takes into

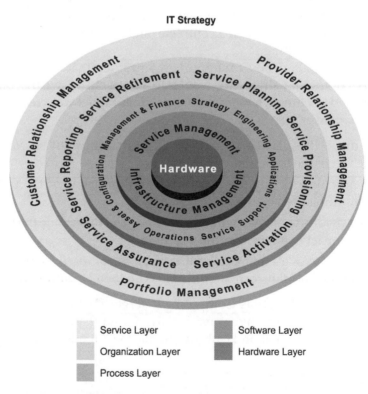

Figure 7.1 The utility computing reference model.

consideration the following elements of a utility computing service:

- business relationships – the *service* layer;
- people and skills – the *organization* layer;
- activities and controls – the *process* layer;
- software implications / requirements – the *software* layer;
- hardware implications / requirements – the *hardware* layer.

For the purposes of describing the model in this section of the book, we will decompose the outer three of these layers into a series of constituent components and give a full description of each. The 'technology' layers (software and hardware) are discussed within Chapter 10 of this book.

7.2 THE SERVICE LAYER

The *service* layer of the reference model defines how abstracted utility computing services are obtained from the service provider, managed and delivered to the service consumer. The associated activities are defined in the 'process layer' and are required to be performed in accordance with the defined processes and to the agreed quality of service parameters.

It is envisaged that this layer will be implemented as an organization entity, which initially may lie 'outside' existing functional areas in the 'organization layer' in an overlay manner (rather than replacing / duplicating those functional areas).

The 'service' layer of the reference model defines how the utility is managed as a business entity, with emphasis on all aspects of the consumer and provider relationships (See Figure 7.2), including:

- consumer relationship management;
- provider relationship management;
- portfolio management.

Each of these elements of the 'service' layer is described, in detail, in the sections that follow.

7.2.1 Consumer relationship management

Utilities are all about 'the consumer'. Without consumers who subscribe to, and make use of, utility services, the utility will not survive.

Figure 7.2 Customer service: it is more than just being on the end of a phone.

It is often assumed that an IT utility that does not have revenue generation as a goal (i.e. a 'value center', rather than a 'profit center') can take consumer relationship management less seriously than a utility that needs to generate profit. In our experience, this is simply not the case. Delivering IT services through a utility model involves having to compel consumers to interact with the service very differently. If significant effort is not focused on ensuring that consumers are happy with the utility model and on taking constructive feedback from them as to how it could be improved, it is likely that the consumers will choose an alternative (perhaps insisting that they continue to be provided with a more traditional method of service delivery, or even moving towards outsourcing the service entirely to an external service provider).

To manage consumer relationships correctly, we believe that effort and focus needs to be given to 11 key disciplines. In this section we describe the rationale behind the inclusion of these disciplines within our reference model. We will also explore some specific considerations that should be given to the disciplines within the context of building an IT utility. In summary, the 11 key disciplines that constitute an effective consumer relationship management method are:

- sales;
- marketing;

- communications/notifications;
- service definition;
- consumer interface;
- service delivery measurement;
- usage reporting;
- support;
- entitlement;
- billing;
- consumer management.

Let us now explore each of the key disciplines in this part of the model (See Figure 7.3) so as to make clear the rationale and considerations of each. Table 7.1 allows quick reference to the attributes of the utility computing Reference Model.

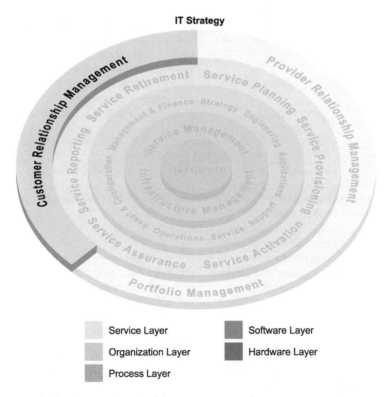

Figure 7.3 Customer relationship management.

Table 7.1 Consumer relationship management disciplines

Sales

Rationale	Considerations
A key aspect of operating a utility computing model is offering a compelling range of IT services. Potential and actual customers for these services will require access to commercial and functional details regarding the nature of the service, the difference(s) between various levels (classes) of service offered, the method of subscription and usage pricing (if appropriate).	Utility services should provide a sales facility for the various services offered. This will include maintaining up-to-date pricing information for each of the services offered with details of relevant operational factors (service availability, service quality). This function includes responsibility for maintaining an accurate and up-to-date customer list for each service offered.

Marketing

Rationale	Considerations
Successfully operating a utility computing service involves compelling potential consumers to make use of the utility service facility. This means making these potential users aware of, and interested in, the range of IT services available.	Utility services should provide a marketing facility for the various services offered. This will include maintaining an up-to-date description of each of the various services offered with details of service availability (launch date, retirement date). This function includes responsibility for promotional activities and maintaining accurate records of prospective customers for existing and future services.

Communications/notifications

Rationale	Considerations
A key differentiator in operating a utility computing model is effectiveness of communications. A variety of communications channels (e.g. email, website, SMS, post, etc.) will be required to inform customers and prospects in a timely and efficient manner, depending upon their preferences and the nature of the communication. The area of communications is of particular importance during the initial launch of the utility service. Effective corporate communication can ensure that all business stakeholders are fully aware and supportive of the new IT delivery model.	Utility services should provide the appropriate communications channels in compliance with any applicable regulations (such as the Data Protection Act). Actual and potential customers should be able to select their preferred channel for formal notification (changes to service terms and conditions, for example). Effective management reporting, showing the effectiveness of the IT utility, will be critical in ensuring that the senior executives of an organization are supportive of the utility service throughout its life.

Table 7.1   (*Continued*)

Service definition

Rationale	Considerations
Adopting a utility computing model requires clear and unambiguous definitions of each service offered. Each definition should be 'necessary and sufficient' for a customer to determine if that particular service is appropriate to his/her needs.	Utility services should maintain an accurate catalog of the services offered. A comprehensive description of each service should be provided, including the key performance areas that differentiate the service from any other. Typical key performance areas might include: time to provision the service, hours or frequency of operation, mean time between failures, recovery time objective, or similar.

Consumer interface

Rationale	Considerations
Potential and actual customers will need to raise service requests and obtain access to service information and contractual data (e.g. cost of service, service usage, etc.). This type of information should be made accessible in a format and through a channel that is reliable, secure, flexible and efficient (from the perspective of the customer).	Utility services should provide a customer portal, through which all types of service information and requests may be obtained. This could take the form of a web-based portal, a call center, or some other process-driven mechanism (e.g. email request/response). Key attributes for the customer interface are ease of use, reliability, consistency (in terms of ergonomics and performance) and range of facilities offered (e.g. service request, service reporting, etc.). A 'how to' guide should be published to inform customers of the facilities provided for each of the key functions they are likely to perform.

Service delivery measurement

Rationale	Considerations
Meeting and exceeding the delivered services key performance areas is a key objective for utility services. The appropriate mechanisms for monitoring and reporting of contracted and actual delivered performance must be implemented and maintained.	Utility services should ensure that all the service key performance areas are proactively monitored and reported upon in a transparent manner. Service performance data should be reported on in a concise, relevant and timely manner consistent with the service definition. Automatic escalation of service violations must trigger the appropriate policy.

(*Continued*)

Table 7.1 (*Continued*)

Usage reporting	
Rationale	**Considerations**
It is the responsibility of the utility service to maintain accurate records of all service requests and all services consumed. Service requests should be authenticated before the request is executed, using the defined policy(s).	The utility service should ensure the appropriate mechanisms are maintained to monitor accurately all requests for service, to establish the legitimacy/entitlement for each request and the outcome of performing each request. The information set should include the identity of the user/requestor, service and class requested, outcome of authentication, date/time of service provided, service quality/compliance indication, service outcome (status), etc.

Support	
Rationale	**Considerations**
Irrespective of the primary consumer interface (see above), a support function must be available for each service offered. The support mechanism (method of accessing support) must be defined clearly, together with the quality of service standards that govern the provision of support.	The utility service should ensure that the appropriate support mechanisms are made available, and that support service levels meet, or exceed, the stated quality of service parameters. Support for portal-based consumer interfaces should include the facility for the consumer to interact with a 'warm body' should the need arise. Depending on the scope/reach of the service and the customer base, multilingual support may be desirable and/or necessary.

Entitlement	
Rationale	**Considerations**
The adoption of a utility service model offering differentiated levels of service has an implication in terms of entitlement to service. Prior to accepting a request for service, the request must be validated in conformance with the appropriate policies and the service request declined or accepted accordingly.	The utility service should ensure that the appropriate mechanisms and supporting data are maintained to ensure that requests for service may be validated properly. The type of validation may depend upon the charging model adopted for the particular service – subscription, usage or combined – and the level of service being requested. Customers requesting a service should be offered timely feedback as to the outcome of their requests.

Table 7.1 (*Continued*)

Billing

Rationale	Considerations
Billing (i.e. charging, whether actual or notional) for services is a key principle of a utility computing model. Without billing, there is no mechanism to evidence the value of services delivered. The utility service should maintain (as noted above) accurate records of service consumption by type and class of service.	Service usage should be reported to the customer on a periodic or ad hoc basis (as defined in the service definition) with associated billing information. The billing data may also be used to update customer records for entitlement checking (for example balance remaining for prepaid services). Billing data will include credits (i.e. service provider penalties) for any services delivered that were not in compliance with documented quality of service parameters.

Consumer management

Rationale	Considerations
One of the objectives of a utility computing model is to reduce IT cost of ownership. Incumbent in this objective is a responsibility to assist users to make the most efficient use of IT resources and facilities. This is achieved through proactive consumer management.	The utility service should include an account management function to monitor service usage and charges, and to review usage patterns. This function should offer advice and guidance (best practice) to encourage customers to reduce their IT costs whilst continuing to meet serviceability, performance and risk criteria.

7.2.2 Provider relationship management

A utility service's relationships with its providers (hardware vendors, distributors, managed services companies, etc.) are absolutely critical to the successful operation of the utility (See Figure 7.4). Part of the inherent benefit of the utility model in IT is that provider/consumer relations can be transformed radically (for the better) right across the IT 'supply chain'. The transformation of the IT department's (utility service provider's) relationship with the IT consumer is reasonably well known (and discussed at length in this book), but it is also important to recognize that the relationship between the IT department and its suppliers can be improved fundamentally as a result of the adoption of utility computing. Our experience has shown us that IT departments that adopt the utility model can introduce an increased amount of flexibility and efficiency into

Figure 7.4 Liaising with providers: How can they help you?

the procurement and supply of IT assets. This is due largely to the fact that, using a utility model, the management of technical capacity can be far more accurate than in traditional IT models.

There are really two main areas that have to be modeled carefully and managed when IT utilities are deployed relating to provider relationship management. These concern:

• the ability to source (procure) IT assets in a timely and appropriate fashion;

• the ability to report accurately on service usage so that resource capacity can be optimized.

For the purposes of this chapter we will refer to the provider of the IT utility service as the *utility service provider* (or USP) and the provider of IT assets to the IT utility service (for example, vendors or managed service providers) as the *IT service provider* (or ITSP).

To manage IT service provider (ITSP) relationships correctly, we believe that effort and focus needs to be given to six key disciplines.

In this section we describe the rationale behind the inclusion of these disciplines within our reference model. We will also explore some specific considerations that should be given to the disciplines within the context of building an IT utility. In summary, the six key disciplines that constitute an effective provider relationship management method are:

- provider service definition;
- procurement;
- delivery standards;
- performance measurement;
- usage and performance reporting;
- accounting.

Let us now explore each of the key disciplines in this part of the model (See Figure 7.5) so as to make clear the rationale and considerations of each (See Table 7.2).

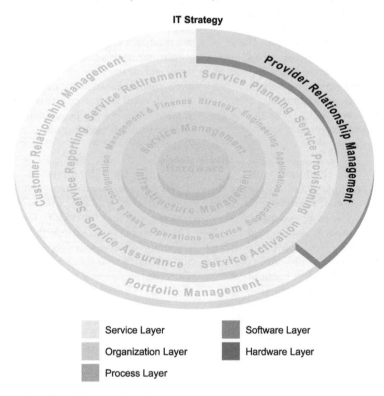

Figure 7.5 Provider relationship management.

Table 7.2 Provider relationship management disciplines

Provider service definition

Rationale	Considerations
Adopting a utility computing model introduces a sense of decomposition of the IT supply chain and differentiation of the roles of service provision, service consumption and service management. The service definition for the provider of raw IT assets (servers, storage, etc.) is a technical description of the service to be offered that may incorporate elements of architecture and specification within it.	Utility service providers should maintain an accurate definition of the service to be provided, together with any material architecture requirements deemed necessary to deploy or provision that service. The definition should include key performance areas that the IT service provider is obligated to meet, with clearly documented penalties for failure to comply.

Procurement

Rationale	Considerations
Planning to implement a defined utility service involves establishing clear requirements that include estimates of demand (for capacity estimation) and quality (for architecture consideration). These parameters will provide a baseline for evaluation of alternative suppliers as the Utility Computing Reference Model does not assume that each utility service will be provided by the incumbent supplier.	The utility service provider should create a set of service provider key performance areas that are mapped onto (with gap analysis) the customer key performance areas to ensure adequate coverage. The provider key performance areas, are likely to be a 'superset' of the customer key performance areas, as the utility service provider will wish to ensure that the quality of service can be maintained during multiple simultaneous usage.

Delivery standards

Rationale	Considerations
A key principle of a utility service is the provision of a service delivered in accordance with agreed quality of service standards at an agreed cost. It is, therefore, essential to establish clear, unambiguous and provable criteria to which the delivery of the utility service must conform. These criteria could include availability and performance targets, number and duration of outages, mean time to restore service, etc.	The utility service provider should create and agree a 'service management' process with the IT service provider. This process should include the definition of service delivery standards, the mechanisms through which the delivered service will be monitored, methods to be used for routine reporting and review, notification and escalation processes, etc.

Table 7.2 (*Continued*)

Performance measurement

Rationale	Considerations
Ongoing monitoring and measurement of service delivery performance is essential to ensure conformance with documented delivery standards. Sufficient data should be captured and retained to enable the appropriate performance reports to be produced and to provide a fact base for assuring the quality of the delivered IT service actually achieved and for instigating the appropriate remedial actions (as/when required).	Utility service providers should work in conjunction with the IT service provider to ensure the appropriate processes and mechanisms to monitor and record delivered service quality are available and tested. Utility service providers should drive the reporting and review processes, and should initiate any required notification/escalation action accordingly.

Usage and performance reporting

Rationale	Considerations
Ongoing measurement and reporting of service consumption data ('metering') is required to facilitate chargeback and trend analysis for demand forecasting. Reporting on the performance of delivered services is necessary to evidence conformance (or lack thereof) with agreed service levels.	Utility service providers should ensure that records of service usage and performance measurements taken during the provision of services are retained for ad hoc and period reporting. Portal-based enquiry and reporting facilities should be made available for 'self service' by consumers and provider personnel.

Accounting

Rationale	Considerations
In operating a utility services model, it is necessary to establish and monitor the financial viability of the service on an ongoing basis. This requires the maintenance of usage and chargeback records to evidence the financial status of the service and to validate ROI projections and demand/capacity forecasts. Periodic management reporting of this data should be undertaken as a matter of routine.	Utility service providers should design and implement the appropriate financial reports for each service in the portfolio, and should ensure that the required date records are captured and maintained to allow routine and ad hoc production of a P&L summary for each service. This type of reporting should be performed irrespective of whether chargeback is notional or actual, to facilitate future investment/de-investment decisions.

7.2.3   Portfolio management

Our experience has shown us that IT services that are to be delivered using the utility model should be managed tightly and managed as a portfolio. The portfolio of services offered will become the most substantial attraction for potential subscribers and will act (when it is constructed well) as a key differentiator from the utility service provider's internal or external competition. Some IT utilities will be designed with many different services made available to consumers; others will have very few. In all cases, it is vitally important that the services that are offered are managed effectively (See Figure 7.6). Managing utility services as a portfolio should ensure that services are:

- always relevant and appropriate to the consumer population for which they were designed;
- offered with enough differing service classes so that all consumers can subscribe to an appropriate level of service;
- priced appropriately so that (where chargeback is in place) the service offerings remain compelling but can generate an appropriate amount of revenue;

Figure 7.6 Managing the portfolio: critical to improving service and reducing costs.

- launched and retired at appropriate times and in a way that minimizes business risk and maximizes service acceptance and subscription.

We have witnessed many IT departments that create, what are initially, very compelling service portfolios as part of a new IT utility, but who then take their eye off the ball and do not manage proactively the portfolio of service offerings as time goes by. This is often a big mistake. Consumer needs and desires change very quickly. The IT industry is very good at exciting technology users with promises of new and compelling offerings, and an IT utility that does not seek proactively to understand the needs and desires of its existing, and potential, consumers can find itself competing aggressively to keep its existing subscribers, quite apart from attracting any new ones!

To manage an IT utility services portfolio correctly, we believe that effort and focus needs to be given to ten key disciplines. In this section we describe the rationale behind the inclusion of these disciplines within our reference model. We will also explore some specific considerations that should be given to the disciplines within the context of building an IT utility. In summary, the ten key disciplines that constitute an effective portfolio management method are:

- service planning;
- class of service modeling;
- service definition;
- service costing;
- provider selection;
- service launch;
- service business review;
- service redefinition;
- service retirement;
- demand forecasting.

Let us now explore each of the key disciplines in this part of the model (See Figure 7.7) so as to make clear the rationale and considerations of each (See Table 7.3).

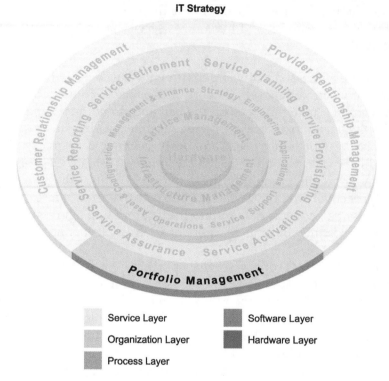

Figure 7.7   Portfolio management.

Table 7.3   Portfolio management disciplines

Service planning	
Rationale	**Considerations**
The development of a portfolio of IT utility services should be viewed as an incremental, evolutionary undertaking. Each service should be evaluated in terms of the business and IT priorities, the likely complexity and cost to introduce and the benefits to IT and the business that would accrue from its inclusion in the portfolio.	Utility service providers should perform a calibration exercise (assessment) to identify and prioritize target areas for utility transformation projects. This exercise should be a collaborative engagement that involves the business community as well as IT personnel. Having agreed on the target utilities, a business case should be prepared to establish the current (incurred) TCO and the anticipated future TCO to support ROI calculations that justify the investment demanded.

Table 7.3 *(Continued)*

Class of service modeling

Rationale	Considerations
For each target utility service it is essential to define the scope of the overall service and the classes (levels) of service to be offered. This requires the establishment of meaningful parameters (from a consumer perspective) that differentiate clearly between the different service classes. Key performance areas that could represent such parameters include time to provision, response time, availability, recovery time objective, etc.	A utility service provider is responsible for ensuring that class of service modeling is performed in a collaborative manner, involving a representative group of consumers as well as IT personnel. A mapping exercise should then be performed to create an architecture for the underlying infrastructure that is capable of supporting the SLO for that particular class, with a corresponding services workflow.

Service definition

Rationale	Considerations
To manage the utility, two separate service definitions should be created – one for the service consumer and one for the service provider. These service definitions should be sufficiently detailed to support unambiguous and sustainable service level agreements – working on the principle that 'you can't manage what you don't measure' – that can be operated in 'back-to-back' mode by utility services.	The utility service provider should create a process for the lifecycle of defining, formalizing, monitoring and maintaining the service level agreements with the service consumer and the service provider. These SLAs should reference the SLO established with both groups and should define the notification and escalation procedures.

Service costing

Rationale	Considerations
Many businesses do not run IT 'as a business', which constrains the ability of the IT organization to position itself as anything other than a cost center. Adopting the principles of utility computing requires the creation of consumer and provider cost models for each class of utility service. The data collected will help to provide transparency of financial performance, and hence enable IT to maintain a services 'chart of accounts'.	The utility service provider should create a cost model to capture one-time (setup) and recurring costs for service provision and consumption. A services 'chart of accounts' should be maintained to evidence the financial performance of the service from both provider and consumer perspectives. The cost of provision should be compared periodically with the envisaged TCO, and a current ROI computed for comparison with the original (used to justify the implementation of the utility).

(Continued)

Table 7.3 (Continued)

Provider selection

Rationale	Considerations
In most organizations, the IT organization functions as both the service provider and the service manager (i.e. it performs the role of utility service provider). This can cause a conflict of interest within IT and a lack of alignment between IT and the business. In the Utility Computing Reference Model, the IT 'supply chain' is deconstructed and the role of service manager segregated from that of services provider.	When a complete service definition is available, the utility service provider should solicit proposals from the internal IT organization and external providers, as appropriate. This creates an effective benchmark from which to assess the capabilities, costs and commitment proffered by the IT organization, whilst providing a clear commercial platform for supplier selection and negotiation.

Service launch

Rationale	Considerations
The introduction of a utility service requires careful planning and communication. For many organizations, the availability of 'self-service' IT facilities on a 'pay-as-you-use' basis will represent a radical departure from 'normal' IT services. The launch of a new service represents a marketing opportunity for the IT department and should be viewed in that light.	The utility service provider should consider the launch of a new utility as a 'marketing' opportunity to enhance and extend the perception of IT and its brand value. A formal launch program should be constructed with the assistance of marketing specialists. This program should focus on communicating the scope, rationale and availability of the new service and should help to ensure that the maximum potential for positive impact is achieved fully.

Service business review

Rationale	Considerations
In a utility model, IT service has a lifecycle that extends beyond design and deployment. The ongoing alignment of the service with business requirements is an essential component of the Utility Computing Reference Model. Each utility service should be subject to periodic formal review using consumption, performance and chart-of-accounts financial data, to create an objective assessment from which to take the appropriate action.	The utility service provider should create and publish a service review process to involve consumers, the IT service provider(s) and the utility service provider. Data captured from service monitoring (consumption rates/trends, delivery performance, costs and chargebacks, etc.) should be analyzed and published in advance of the meeting. The review should focus on establishing to what extent the current service meets the business requirements, and what (if any) adjustments are necessary.

Table 7.3 (*Continued*)

Service redefinition

Rationale	Considerations
One possible outcome of the service business review is that the service (or a particular class of service) no longer meets the requirements of the business and/or the consumers. In such a case, the terms of service definition should be adjusted to bring the service/service class into line with the agreed requirements, and the corresponding changes should be formalized.	The utility service provider should initiate a service redefinition program to realign the service/service class with business requirements. This may involve changes to the requesting/provisioning process, adjusting service delivery SLO, or similar. Any changes to the service definition should be communicated properly to the consumers in advance of the change being implemented.

Service retirement

Rationale	Considerations
One possible outcome of the service business review is that demand for the particular service, or service class, is insufficient to warrant ongoing provision. In such a case, an impact assessment must be performed to establish what action may be necessary to offer an alternative service, protect/migrate stored data, change business processes, etc. prior to retiring the service.	The utility service provider should define a service retirement program, triggered from the service review/redefinition process. The process should formalize the activities associated with service retirement, including those mentioned left, but also considering the decommissioning of equipment/facilities and the phasing in of any alternative/replacement service.

Demand forecasting

Rationale	Considerations
Another possible outcome of the service business review is that the profile of demand for the particular service, or service class, is changing. Consumption trend data should be utilized to inform the service provider of actual/anticipated changes in the expected usage, so that the appropriate actions may be taken in a timely manner to assure ongoing service delivery/quality compliance with agreed SLO is not compromised.	The utility service provider should assess service consumption and service delivery performance data on a routine basis and identify changes in the demand profile of each service/service class. Notable changes (either in usage or in delivery performance) should be analyzed to determine the cause. Any requested changes to service provision should be communicated formally to the service provider with a description of the symptoms, causal analysis and proposed action.

7.3 THE PROCESS LAYER

The *Process* layer of the reference model defines the management processes and various activities that are required for the effective and efficient operation (from its inception to its retirement) of the IT utility. These processes can be viewed in the form of an 'IT utility lifecycle' (See Figure 7.8).

As we can see, the service lifecycle that we have modeled within the Utility Computing Reference Model consists of six stages that span the entire life of an IT utility. This process model has been designed to map directly to the associated layers of the reference model ('Organization' and 'Service') and it has been modeled, also, to cater for any re-planning (redefinition) of utility services that may need to occur during the life of the IT utility.

This part of the Utility Computing Reference Model has been constructed through a process of synthesizing several industry standards (ITIL and CoBIT to name two) that we thought bore relevance to how an IT utility should be modeled.

As a project team starts to strategize and plan for utility transformation, it is important that they consider carefully how the processes that are detailed within this section will be enabled within their existing organization. The relationship between this layer and, in particular, the organizational layer of the reference model, should be examined and designed so that the organizational structures that exist as a result of transformation support fully the various process elements that are described here.

7.3.1 Service planning

The planning of the utility service should be structured in a way that allows the project team to plan and design transformation whilst

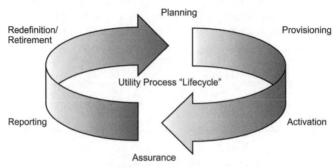

Figure 7.8 Utility process lifecycle.

taking into consideration all elements of the utility service and all stages of the utility service lifecycle. Many of the utility computing plans that we have reviewed focus heavily on initial service characteristics, but do not consider in detail the ongoing running and operation of the utility environment once it has been established fully.

To plan a utility service correctly, we believe that effort and focus needs to be given to six key disciplines. In this section we describe the rationale behind the inclusion of these disciplines within our reference model. We will also explore some specific considerations that should be given to the disciplines within the context of building an IT utility. In summary, the six key disciplines that constitute an effective service planning method are:

- research and requirement analysis;
- service concept modeling;
- resource and cost modeling;
- service delivery management;
- price and service options;
- service economics.

Let us now explore each of the key disciplines in this part of the model (See Figure 7.9) so as to make clear the rationale and considerations of each. Table 7.4 allows quick reference to the attributes of the Utility Computing Reference Model.

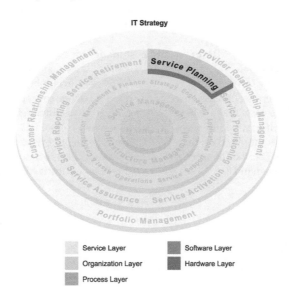

Figure 7.9 Service planning.

Table 7.4 Service planning disciplines

Research and requirement analysis

Rationale	Considerations
The building of a comprehensive portfolio of utility services requires an ongoing commitment to monitoring, establishing and validating requirements and/or opportunities for such services.	Consider creating a 'service opportunity' process to capture in a database all requests for IT services that cannot be satisfied by the existing portfolio. Allow entries based on business need, research and personal opinion. Institute a regular review process and provide feedback to the originators of each entry to cultivate a sense of responsiveness and innovation.

Service concept modeling

Rationale	Considerations
Prior to investing in the development of a new utility service it is essential to evaluate the service opportunity from a variety of dimensions, including the services to be abstracted, the key processes required, the likely/anticipated demand and the technical requirements/feasibility.	Consider defining a workshop process at which the key stakeholders from the organization debate the opportunity for the service in a structured manner: establish the key services to be offered, identify the principal processes (workflows) through which each service will be delivered and determine the key enabling technologies. Capture and validate (as far as practical) low and high watermarks for the expected demand.

Resource and cost modeling

Rationale	Considerations
Establishing a new service has implications in terms of the resources required and the costs that are likely to be incurred. Indicative resources and costs can be computed on the basis of a defined service concept and the anticipated demand.	Consider developing a model for the agreed service concept to capture indicative resource requirements and associated costs – linked, where appropriate, to expected volume or demand. Use the model to provide a 'first-cut' indicative ROI.

Table 7.4 (*Continued*)

Service delivery management

Rationale	Considerations
The effective management of service quality is a key objective when operating a utility service. Service delivery management should be proactive wherever possible to allow sufficient time for corrective action prior to commencement of service degradation.	Consider creating a high-level design showing how each of the SLO for both the service and the service consumer will be monitored, measured and reported. A process should be created for detecting, diagnosing and rectifying QoS defects – this process should include the automatic capture and recording of service violations.

Price and service options

Rationale	Considerations
It is likely that a number of service variants may be required in order to provide effective choice to the consumer. Each variant should be differentiated by having unique values and/or non-overlapping ranges for the consumer SLO, together with its own pricing schedule.	Consider creating a template based on 'best practice' to define service variants and prices. Alternatively, conduct a joint IT/business workshop to identify key service stakeholders and use the workshop process to establish the appropriate variants and prices.

Service economics

Rationale	Considerations
Running a utility service should be considered as a business activity, and should be managed as such – irrespective of the pervading attitude to notional or actual chargeback. The concept of IT value cannot be inculcated unless there is visibility for cost.	Consider building a P&L model of exposing the financial viability of the service. The model should be based on variables such as the indicative cost and price calculations. These variables should be parameterized to allow 'what if' scenarios to be modeled and compared. The model should then be capable of being run in 'goal seeking' mode – i.e. to compute a price per service variant based on stated assumptions that delivers a 'break even' P&L.

7.3.2 Service provisioning

In the provisioning of a utility service it is important that all data points captured during the planning phase are taken into consideration, and that all layers of the Utility Computing Reference Model are given appropriate focus and resourcing. In most cases, the provisioning phase of a project will be the first time that a project team is visible to the business in actually deploying technologies and rolling out new processes and structures. Care should be taken, therefore, that this phase is well managed and executed. Potential utility consumers could be put off from subscribing to your new service if your execution of initial provisioning is not seen to be run professionally and well organized !

Our process model breaks the utility provisioning process down into seven key disciplines. In this section we describe the rationale behind the inclusion of these disciplines within our reference model. We will also explore some specific considerations that should be given to the disciplines within the context of building an IT utility. In summary, the seven key disciplines that constitute an effective

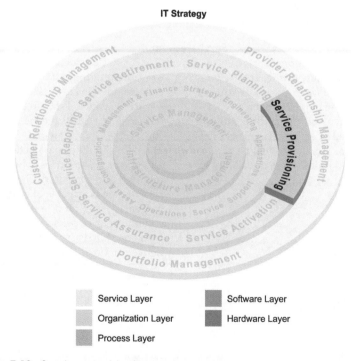

Figure 7.10 Service provisioning.

service provisioning method are:

- service class modeling;
- workflow modeling;
- architecture blueprinting;
- business case validation;
- implementation planning;
- preliminary implementation;
- testing.

Let us now explore each of the key disciplines in this part of the model (See Figure 7.10) so as to make clear the rationale and considerations of each. Table 7.5 allows quick reference to the attributes of the Utility Computing Reference Model.

Table 7.5   Service provisioning disciplines

Service class modeling	
Rationale	**Considerations**
Offering a range of service variants (differentiated classes of service) affords increased alignment between the criticality of the service to the user and the costs associated with providing and using the service (economic delivery).	Consider establishing the minimum number of service variants that can support clearly differentiated services (in terms of consumer SLO). Map those variants onto corresponding SLO for the enabling technologies. For example, a consumer SLO of 'recovery point objective of zero' (i.e. no data loss) could map to a provider SLO of 'synchronous replication'.
Workflow modeling	
Rationale	**Considerations**
The adoption of standard operational processes is key to reducing lifecycle cost of ownership, increasing consistency of action, enabling self-service, collecting consumption data and minimizing latency (time to take action).	Identify key workflows required for service activation, operation and management. Decompose these workflows to identify the key actions and interfaces. Look hard for opportunities to automate actions. Create service level interface (SLI) standards to define the flow of information between the actions. Create service level agreements (SLAs) for the elapsed time permissible to each action, with defined escalation procedures.

(Continued)

Table 7.5 *(Continued)*

Architecture blueprinting

Rationale	Considerations
Achieving consistent QoS within each service class requires the adoption of technology standards. A blueprint for each service class must be created showing the major components and the standards approved for each.	For each technical component required to provision or action the service, research and define the permissible options for component selection that are commensurate with achieving consistently the required SLO. Group these component selections into the appropriate service variants and ratify through the IT governance process to create a comprehensive suite of standards.

Business case validation

Rationale	Considerations
With more detailed service design elements in place, there is now sufficient data to validate/refine certain of the assumptions used to build the 'service economics' P&L. Lifecycle-based cost-of-ownership calculations provide a more complete (and substantive) view of the financial implication of adopting a utility model.	Expand the existing P&L model through 'before' and 'after' TCO analysis. Compute an associated ROI to assure financial viability of implementing the service – this is an essential step when using 'cost savings' (from current state) to drive a transformation from an existing IT model to utility.

Implementation planning

Rationale	Considerations
A project to implement a utility computing model represents a transformation that extends beyond a pure technology play. A formal approach to implementing the utility service is vital as the 'reach' of the project will involve and impact many different stakeholders.	Consider planning the implementation from the perspectives of each of the various stakeholder groups and adopting a workshop format to create an understanding of the transformation and cultivate a willingness to embrace change. The success of a transformation project is contingent upon robust/realistic planning, early involvement/continued consultation and frequency/transparency of communication.

Table 7.5 (*Continued*)

Preliminary implementation

Rationale	Considerations
Implementing the utility service requires all aspects of the plan to be actioned – with focus on control, communication and coordination. A project office should manage the different aspects of the transformation and an executive sponsor with the appropriate authority should act as the director and arbitrator.	Consider having four separate teams within the project office to 'own' particular aspects of the implementation. One team should focus on business/organization activities (including communications), another on process and controls, a third on technology deployment, and the fourth on integration.

Testing

Rationale	Considerations
Prior to service launch, the newly constructed (or modified) utility service should be tested thoroughly. Testing should include functional validation for each service variant, together with SLO monitoring and compliance certification. Operational process and management controls should also be inspected and tested.	Consider developing a testing strategy based on comprehensive coverage within each of the four teams, with an additional 'end-to-end', or integrated, test suite. The utility service testing should assure the performance of each service variant, verify QoS monitoring and reporting, confirm accuracy of usage data collection, and validate that the management and operational controls and procedures function as designed.

7.3.3 Service activation

Having provisioned all aspects (technical and otherwise) of the new IT utility, it will be time to activate the service and make it available to consumers. This phase of a transformational project is obviously very visible to the potential consumer (and their management teams – probably your business sponsors!); as such it is vital that the same degree of professionalism and efficiency is shown here as was displayed in the previous project phases.

This part of the process lifecycle includes a 'service launch' process. The activities undertaken here will be vital to the first impressions of the new service that are conveyed to potential consumers. In focusing on this particular phase, organizational culture and communications should play a major role in your plans (refer to Chapter 11).

To manage the launch and activation of an IT utility, we believe that effort and focus needs to be given to six key disciplines, these are the following:

- service planning;
- service launch;
- resource allocation;
- service activation;
- consumer administration;
- user training.

It cannot be stressed enough that this particular part of the utility process lifecycle can absolutely make or break a project. If the 'planning' and 'provisioning' phases have gone well, and their objectives met, then a project team should be in good shape at this point. However, a badly executed 'activation' phase could easily

Figure 7.11 Service activation.

undo all of the good work that has gone before. Hope for the best and plan for the worst at this point in the process lifecycle. You will never have enough people to support the utility service launch, take help wherever you can get it and do not let your project team rest on their laurels here.

Let us now explore each of the key disciplines in this part of the model (See Figure 7.11) so as to make clear the rationale and considerations of each. Table 7.6 allows quick reference to the attributes of the Utility Computing Reference Model.

Table 7.6 Service activation disciplines

Service launch	
Rationale	**Considerations**
Offering key infrastructure facilities in the form of utility services requires the adoption of a more business-oriented approach to customer communications. The launch of a new service provides an excellent opportunity for the ITO to adopt more vigorous and visible communications channels and vehicles.	Consider using an external agency to develop a multimedia launch program (email, bulletin, website). Integrate these forms of communication with an ongoing commitment to customer relationship management.

Resource allocation	
Rationale	**Considerations**
Certain types of utility service may require components to be made available in a timely manner to assure continuity of service. For example, a server utility may not be populated fully at implementation time with sufficient numbers of each server type.	Consider updating the demand forecast assumptions used for planning purposes with subscriber data to verify whether or not sufficient resources are available. Take the appropriate action to minimize the likelihood of resource shortfall and consequent service outage or degradation.

Service activation	
Rationale	**Considerations**
The utility service should be made available officially to consumers in a structured and heavily governed manner. This phase can be used to validate user entitlement to service and activate/decline the request accordingly.	Consider activating the utility service to one population of users at a time (under the guidance of a published timetable). Start with the 'friendly users' so that any issues with the service will only be experienced by consumers of your choosing.

(Continued)

Table 7.6 (*Continued*)

<table>
<tr><td colspan="2" align="center">Consumer administration</td></tr>
<tr><td align="center">Rationale</td><td align="center">Considerations</td></tr>
<tr><td>At utility launch it will be necessary to create and maintain the appropriate user entitlement/activation process/ controls. These will feed the 'business interface' function, detailed in the organizational layer of the reference model.</td><td>Consider carefully how consumer entitlements and privileges will be administrated and, in particular, authorized. Interfaces will need to be built with business managers so that they can authorize appropriate levels of access to the utility for their staff.</td></tr>
<tr><td colspan="2" align="center">User training</td></tr>
<tr><td align="center">Rationale</td><td align="center">Considerations</td></tr>
<tr><td>At utility launch it will be necessary to ensure that any additional skills that are required by potential users of the IT utility are catered for as part of a comprehensive and focused end user training program.</td><td>Train potential users of the new utility appropriately. Our experience is that lengthy, formal classroom training is often not appropriate for this type of exercise. Often, it is worth considering the use of web-based training methods, which allow end users to become as skilled in the use of the new utility as is appropriate to their job function.</td></tr>
</table>

7.3.4 Service assurance

So, the panic is over! The utility service has been launched successfully and any initial problems have been ironed out. The utility process lifecycle, at this point, changes from focusing primarily on issues concerning building and launching the IT utility, to focusing heavily on the ongoing assurance of service levels and consumer satisfaction.

As we have mentioned elsewhere in this book, utilities (of any kind) are exposed constantly to pressures of competition (from other utilities). Service assurance is, therefore, an important part of the process involved in managing a successful and sustainable IT utility. Once again, in creating this part of the Utility Computing Reference Model, we have synthesized several common industry best practices in order to create a process model that satisfies the service assurance requirement of an IT utility.

To manage the ongoing assurance of IT utility services, we believe that effort and focus needs to be given to six key disciplines. These are the following:

- usage monitoring;
- service delivery performance monitoring;
- problem management
- service delivery management;
- service violation/escalation;
- service support.

These process disciplines should operate together to ensure the constant and proactive assurance of service and, in turn, consumer satisfaction.

Let us now explore each of the key disciplines in this part of the model (See Figure 7.12) so as to make clear the rationale and considerations of each. Table 7.7 allows quick reference to the attribute of the Utility Computing Reference Model.

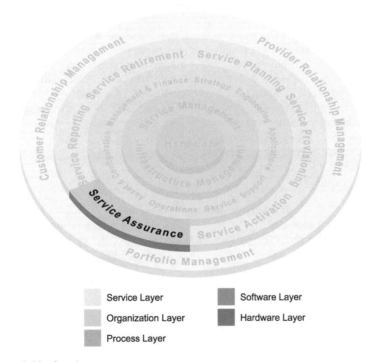

Figure 7.12 Service assurance.

Table 7.7 Service assurance disciplines

Usage monitoring

Rationale	Considerations
The utility service provider will need to ensure timely and accurate monitoring of service usage and update the appropriate service databases on a transaction basis. Any form of financial chargeback mechanism will rely heavily on this.	Select software tools that suit this requirement carefully. Focus on what really needs to be monitored, in terms of usage, and do not over-engineer your monitoring solution. This can lead to increased complexity and operational costs.

Service delivery performance monitoring

Rationale	Considerations
The utility service provider must ensure timely and accurate monitoring of QoS data, and update the appropriate service databases with performance data on a transaction basis. This is critical in measuring a utility service against SLAs and other objectives.	Consider the use of transaction performance measurement tools that are capable of tracking transactions from 'end to end' and that can give detailed root cause analysis information where performance problems are apparent.

Problem management

Rationale	Considerations
The utility service provider must be able to be both reactive and proactive in the capture and resolution of consumer issues. A problem management process should be capable of targeting issues that are either technical, personnel- or process-related. The system should also be capable of creating reports that detail performance associated with problem resolution.	Our experience of implementing structured problem management frameworks has shown us that the Infrastructure Technology Information Library (ITIL) is a source of excellent best practice in this particular area.

Service delivery management

Rationale	Considerations
The utility service provider must maintain appropriate controls for managing service delivery throughout the execution of the service. This should include near-time/real-time alerting to QoS problems.	Good service delivery management is all about good process. Once again, the ITIL standards form an excellent reference for best practice in this area.

Table 7.7   (*Continued*)

Service violation/escalation

Rationale	Considerations
The utility service provider should define the process for notification and escalation of QoS violations and activate it as appropriate. Accurate records should also be stored with supporting transaction-level data to facilitate diagnosis and remedial actions and to evidence lack of compliance with service SLO.	Consider adding this type of data to the business-level reports that are published to focus on the performance of the utility service. Sharing this type of information will show that the utility service provider takes violation seriously and acts in an appropriate way to these types of event.

Service support

Rationale	Considerations
The utility service provider must define the support process and provide the appropriate facilities/ resources to handle support requests and track performance.	Once again, our experience of implementing structured service support frameworks has shown us that the Infrastructure Technology Information Library (ITIL) is a source of excellent best practice in this particular area.

7.3.5   Service reporting

The correct data gathering and reporting of all pertinent and interesting data associated with the IT utility will be crucial, both for the utility service provider (to ensure that they are working constantly with correct information concerning the efficiencies and effectiveness of the utility) and the consumer (so they can be confident that service levels are being adhered to and that they are being provided with the service for which they are paying !). We felt it necessary, therefore, to dedicate an element of the process layer of the reference model to the tasks and competencies that are necessary to sustain this area and ensure that service reporting is adequate for all of the utility computing project stakeholders. It is perfectly possible for a very well-designed and maintained utility to be seen as unsuccessful (and, as a result, not compelling) by current and potential consumers, where service reporting is not designed and run effectively. In the case of many IT utilities, the actual 'service' that is provided (system backups, for example) remains fairly transparent to the consumers. In these cases, in particular, it is vitally important

that the consumer is reminded constantly of the good work that
their utility service is providing to them!

To manage the effective reporting for an IT utility, we believe that
effort and focus needs to be given to seven key disciplines. These
are the following:

- usage reporting (consumer);
- performance and compliance reporting (consumer);
- consumer billing;
- consumer satisfaction;
- usage reporting (provider);
- performance and compliance reporting (provider);
- utility service profit and loss (P&L) reporting.

Let us now explore each of the key disciplines in this part of the
model (See Figure 7.13) so as to make clear the rationale and considerations of each. Table 7.8 allows quick reference to the attributes of
the Utility Computing Reference Model.

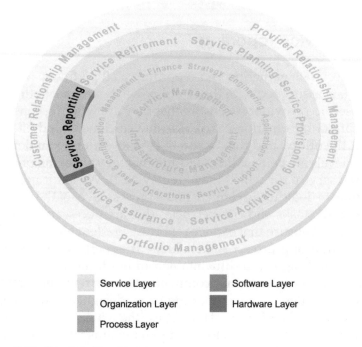

Figure 7.13 Service reporting.

Table 7.8 Service reporting disciplines

Usage reporting (consumer)

Rationale	Considerations
The utility service provider will need to provide general and regular reports on usage data on a 'per service activation' basis. This data will need to be aggregated on a periodic basis so that trend reports can be created.	Consider evolving this reporting over time and storing this data in a relational database that can be queried on an ad hoc basis.

Performance and compliance reporting (consumer)

Rationale	Considerations
The utility service provider will need to provide reports that show QoS characteristics compared against appropriate SLAs. Violation reports should be administered separately and should include detailed causal analysis.	Consider evolving this reporting over time and storing this data in a relational database that can be queried on an ad hoc basis. Also, ensure that your method of reporting can cater for trend analysis from its inception.

Consumer billing

Rationale	Considerations
The utility service provider must be able to provide detailed, accurate and periodic billing data based on usage analysis and compute costs. This billing information will need to be correlated constantly against agreed pricing that has been published to the consumer.	Ensure that any agreed problems that carry financial penalties are recognized as part of this reporting process. Attempting to adjust consumer bills later, in order to take penalties into consideration, will be time consuming and will damage consumer confidence in your utility service.

Consumer satisfaction

Rationale	Considerations
The utility service provider will need to create both formal and informal mechanisms to capture consumer satisfaction levels over time and at regular (and agreed) periods. Consumer satisfaction reports should then be published to all stakeholders on a regular basis.	Consider spending time with business management representatives early on to establish what they would like to see within consumer satisfaction reports.

(Continued)

Table 7.8 (*Continued*)

Usage reporting (provider)

Rationale	Considerations
The utility service provider will need to create general and regular reports on usage data for their own purposes. These reports will focus 'inwards' on how consumer demand relates to compute and human resource requirements.	Consider evolving this reporting over time and storing this data in a relational database that can be queried on an ad hoc basis.

Performance and compliance reporting (provider)

Rationale	Considerations
The utility service provider will need to generate its own reports that show QoS characteristics compared against appropriate SLAs that are held between the service provider and its suppliers (vendors, outsourcers, etc.).	Use these reports as a constant source of debate and negotiation with suppliers in order to ensure that they are providing your business with the best possible levels of service and pricing.

Utility service profit and loss reporting (P&L)

Rationale	Considerations
The utility service provider must define and maintain a detailed profit and loss report that provides accurate records of all financial elements associated with the utility service. This report should contain actual costs and revenues and should allow the utility service provider to create forecast reports based on current financial data and usage patterns.	Carefully consider how much of your profit and loss information should be exposed to your consumers. In some situations it may be entirely appropriate for this type of report to become 'common knowledge'. In other situations you may not want your consumer to have insight into your profit margins.

7.3.6 Service redefinition and retirement

It is important for the process model used to establish utility services to be iterative. Ultimately, the services provided to consumers will be retired and decommissioned. However, it is likely (in the case of successful utilities) that the services provided will need to be redefined and refined in order to take new trends in consumer demands into consideration, or to modify services in order to smooth out issues that are being experienced, whether by the consumer or by the utility service provider. The service redefinition and retirement component of the process model provides the linkage between the end of the process lifecycle and its beginning.

The process elements contained within this part of the Utility Computing Reference Model have been constructed to ensure that a utility service provider is able to keep its service offerings compelling and differentiated, and their presence as part of the utility model will impact all facets of the organization that manages the utility (as discussed within the section on the organization layer).

To manage the iterative redefinition and retirement of IT utility services, we believe that effort and focus needs to be given to six key disciplines. These are the following:

- requirements monitoring;
- utilization and capacity monitoring;
- demand forecasting;
- service parameter adjustment (consumer);
- service parameter adjustment (provider);
- service parameter change notification.

Let us now explore each of the key disciplines in this part of the model (See Figure 7.14) so as to make clear the rationale and considerations of each. Table 7.9 allows quick reference to the attributes of the Utility Computing Reference Model.

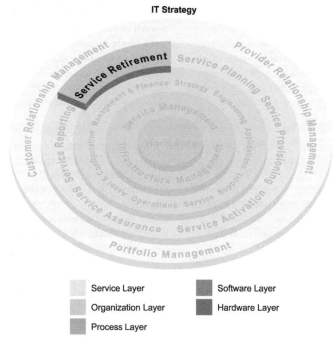

Figure 7.14 Service redefinition and retirement.

Table 7.9 Service redefinition and retirement disciplines

Requirements monitoring

Rationale	Considerations
The utility service provider will need to ensure that service lines and service parameters evolve in line with true (rather than perceived) consumer requirements. A monitoring process should be established that captures baseline data for decision making and facilitates a 'continuous improvement' approach.	Consider setting up a regular meeting with business stakeholders that reviews baseline data and makes decisions regarding the continuous improvement of utility services.

Utilization and capacity monitoring

Rationale	Considerations
The utility service provider must compare and verify consumer and provider usage reports. Planned and committed consumption levels should be studied and possible over/under utilization and consumption risks should be identified.	Consider the use of trend analysis techniques in order to facilitate the possibility of proactive adjustments to technology procurement and consumer commitment levels.

Demand forecasting

Rationale	Considerations
The utility service provider should extrapolate usage data in order to identify likely future demands. This data should be shared with technology providers to ensure that consumer demand does not put service levels at risk through over-utilization of technology assets.	Consider including multiple data points as part of this analysis. These should most likely include periodic (fiscal) variations in demand, growth in customer base and changes to regulations and governance (the Data Protection Act, for example).

Service parameter adjustment (consumer)

Rationale	Considerations
The utility service provider should (on a regular basis) review general usage patterns, consumer satisfaction and service profit and loss (P&L) in order to plan refinement or retirement of utility services.	This regular analysis will prove to be particularly important in the establishment of service pricing characteristics. It will form the basis of decisions made with 'supply and demand' dynamics in mind.

Table 7.9 *(Continued)*

Service parameter adjustment (provider)	
Rationale	**Considerations**
The utility service provider should (on a regular basis) review the service level parameters that exist between themselves and their providers. This will ensure that, as consumer service parameters are adjusted, the supply of technology assets can be adjusted so that it is appropriate.	Consider and discuss the dynamic and potentially unpredictable nature of procurement patterns with suppliers early on, so that a mutually beneficial business relationship with these parties can be established and maintained from the start.

Service parameter change notification	
Rationale	**Considerations**
A formal process is required in order to notify consumers of planned changes to service parameters (availability, cost, performance, and so on). An 'informational' communications channel should also be maintained to provide advice and guidance on how to optimize service usage.	Consider innovative techniques in communication of this type such as the usage of 'notice boards' within a utility service web portal or regular emails based on standard, professional templates.

7.4 THE ORGANIZATIONAL LAYER

The *organizational* layer of the reference model helps to define the type of people (and teams) that will be required to successfully design, build, run and maintain a true IT utility. In creating the reference model, we concluded (based on experience) that to neglect the appropriate transformation of the IT organization during a utility computing project was to put the project at risk from the offset. Not only do some of the traditional IT roles and responsibilities have to change to support the utility model, but some new roles often have to be fulfilled in order for the model to be successful.

When considering the type of IT organization that should be responsible for an IT utility, it is worth taking into consideration both:

- organizational structure and competencies (described within this section); and
- organizational culture transformation (described later in Part 3).

We have found that it is only when these two factors are considered in parallel, during an IT strategy definition, that the organizational ('people') attributes of a utility transformation can be managed effectively.

The reader should bear in mind that the organizational competencies and structures detailed within this section are not designed to be prescriptive. As with the other layers of our reference model, project teams should take the parts of our 'best practice' that are appropriate to their own initiatives and leave other parts alone. Additionally, in this part of a transformation in particular, it is always important to take the current IT organization, its structure, politics, competencies and so on into consideration prior to creating a transformational strategy. Many of the competencies detailed here will already exist within your organization, these should be used where possible. In some areas you may find that existing skills can be extended and broadened in order to satisfy the requirements of your utility computing initiative. This situation can prove to be of great benefit in the context of staff morale and personal growth.

So, what should the IT utility service provider look like? Well, in our opinion, it should be described at the highest level in the form of 'centers of excellence' (See Figure 7.15).

The organizational structures shown here represent all of the fundamental organization functions that are required to manage the complete lifecycle of an IT utility. The reader should not (necessarily) consider these as organization departments or business units.

Figure 7.15 Organizational centers of excellence.

Our experience is that a successful utility service provider may actually be constituted with any of the following types of competence centers:

- departments – where an actual department of individuals takes responsibility for a function;
- virtual teams – where a utility service function is maintained by various departments that are organized to work together across physical, political or departmental boundaries;
- an outsource – where a third party service provider is given responsibility for the function.

As with many areas of utility computing transformation, the selection (or combination) of these approaches to organizational transformation will depend greatly on the culture and nature of the business in question.

Our experience also shows that, often, one center of excellence (whether it be an actual department, virtual team or outsourcer) can take responsibility for two or more of the functions shown here. For example, 'asset and configuration' and 'engineering' functions are often made the responsibility of one department. This, in many cases, is an evolved version of the IT Engineering Department that already existed prior to the utility computing transformation. As another example, we have seen on several occasions that forming a discrete 'strategy' team (as a separate body of individuals who act independently of the rest of the ITO) can cause problems. This type of department can often be seen to form new strategies in isolation from the rest of the IT department and, as a result, can find it very difficult to make their new ideas come to life. The very last thing that this function needs is to be seen as operating in an 'ivory tower' with no linkage or empathy for the real-world of IT operations. The best and most successful 'strategy' functions that we have seen have all been created using the 'virtual team' approach and by selecting thought-leading individuals from all of the other functions that exist across the IT organization. As we mentioned earlier, this organizational model may not be entirely appropriate to every utility computing initiative. Certainly, in our experience of applying this model to an existing IT organization, it is true to say that some of the IT functions shown here tend to take higher priorities than others. Other functions may be addressed initially by a 'virtual team' and will be given dedicated human resources once the size and complexity of the IT utility justifies this additional operational expense. Often, for example, the 'management and financial' function (described in more detail later) will initially be made the responsibility of an existing IT operations group (who already have basic accounting, reporting and business relationship management practices in place). When the IT utility requires more rigorous and comprehensive managerial and financial functions, then the IT organization would recruit additional staff to design, build, execute and maintain these extended functions.

As a general comment on the application of this model we would suggest that, in most instances, the project team would be well advised to recruit additional staff (or obtain help from an outside consulting organization) to focus properly (and with the correct skills) on the initial creation of the 'service' function. This area tends to represent the biggest challenge to the traditional IT organization in terms of transformation since many of its functional attributes are more aligned to customer satisfaction and consumer management than to the technical functions of the utility service.

Now let's explore each of the function centers of excellence in more detail.

7.4.1 The service function

The service function is, arguably, the most fundamentally important function in terms of the consumers' perspective and experience in using the IT utility. Certainly, it is the area of organizational change that often requires the most thought during transformation, since many of its subfunctions will not have been focused on by the ITO prior to the adoption of the utility model. This organizational function primarily holds responsibility for 'the consumer experience' (although, as we will see, it also takes responsibility for the relationships with IT suppliers). It maps directly to the 'service' layer of the Utility Computing Reference Model and its primary job in life is to ensure consumer (customer) satisfaction.

As we can see from Figure 7.16, the service function can be split into five main areas of discipline. Importantly, these disciplines are focused on the management and relationships with both the consumers (the users of the IT utility) and providers (the suppliers of IT assets to the utility service provider).

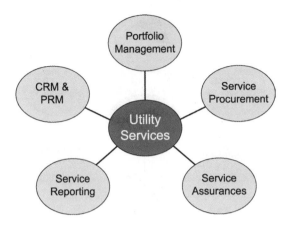

Figure 7.16 IT utility service functions.

As we have mentioned, many typical ITOs initially find the prospect of evolving some of the functional areas fairly daunting ('CRM, from an IT department ?'). It has been our experience, however, that the ITOs that focus heavily on this area tend to be the ones who achieve quick wins with utility computing. This is not least because of the fact that the 'CRM' (consumer relationship management) competence that exists in this model takes a large part in 'selling' the IT utility to its prospective consumers before the service is made available. Additionally, an important role is played by the 'PRM' (provider relationship management) function during the early stages of a utility computing project in forging new and extended commercial relationships with IT vendors.

7.4.2 The operations function

The operations function tends to be fairly akin to most existing ITOs, in that most organizations will have at least basic coverage of the areas illustrated below. This function is responsible primarily for the day-to-day technical maintenance of the assets of which the IT utility is comprised.

The operations function will typically have close linkages to each of the other organizational functions, although it will function day to day with particularly strong ties to the engineering and service functions.

Figure 7.17 IT utility operations functions.

The operations function can be split into six areas of discipline (See Figure 7.17). The reader will recognize that many of the functions traditionally found as part of an ITO are included here (support, systems operations, systems, monitoring, and so on).

Additionally, we have built some functions into the model that may not be found in an existing ITO. These include a specific discipline centered around automation (a center of excellence that specializes in automating the day-to-day management processes that exist to keep the IT estate running in an efficient manner) and a job control function to ensure tight control of scheduled tasks (such as data archiving or system patch distribution and control).

7.4.3   The engineering function

Once again, this function and its disciplines may already exist within an ITO. This function is responsible primarily for the technical implementation and tuning of the IT estate that supports the utility service. As mentioned in the previous section, the engineering function will have close linkages to, in particular, the operations function, and, importantly, a well documented 'hand off' process should exist between these two functional areas to ensure that once engineering have implemented a technical solution, that the people responsible for managing its day-to-day operations are very well briefed as to its technical intricacies and maintenance processes. In our experience,

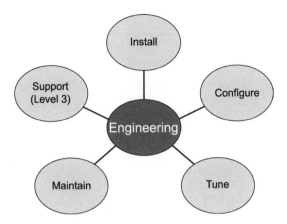

Figure 7.18 IT utility engineering functions.

it is of particular importance that any automated task or process is well understood by the 'automation' and 'job control' functions (within operations) at the point where this center of excellence takes control of a technical asset.

The engineering function can be split into five areas of discipline (See Figure 7.18). The reader will notice that the 'support' function (level 3) is seen here. This competence has direct linkage to the support functions that exist in the other centers of excellence (support level 0 in the support function, described in the next section, support level 1 in the operations function, and so on). It is vitally important that the engineering center of excellence can provide support services to the other organizational functions, since it is within this function that deep technical knowledge and understanding of the IT estate will exist (and can resolve some of the 'deep' technical support issues that may arise in day-to-day utility operations).

7.4.4 The support function

The support function exists as a center of excellence that acts as a conduit to all utility service enquiries (termed *level 0 support* in our model). In our view, this particular function should (from the offset) be staffed by dedicated personnel who are experts in managing 'helpdesk' environments (perhaps based on industry standards such as ITIL).

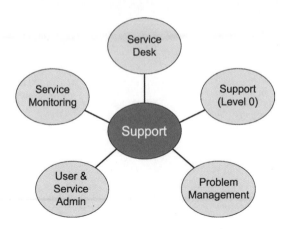

Figure 7.19 IT utility support functions.

Along with the service function (described earlier), this center of excellence is very important in terms of the consumer's perception of the IT utility. No IT service will ever run without issues (technical or otherwise) arising and it is vital that the ITO is seen to manage these issues professionally and efficiently. The project team can responsibly expect that this particular function will be very busy during the initial stages of IT utility launch. There are always a few creases to iron out when a new IT service goes live.

The support function can be split into five areas of discipline (See Figure 7.19). The model shown here owes a lot to the definitions of best practice for service support that are defined within the Information Technology Infrastructure Library (ITIL) standards (described elsewhere in this book). In the IT utilities that we have seen to date, it is fairly typical for a body of 3–5 experts to be dedicated to this function, such is its importance in terms of overall consumer satisfaction.

Effort should be made to create linkages between this function and the CRM discipline (within the service function). Consumer relationships will often be influenced heavily by the method in which consumer queries and problems are managed and resolved.

7.4.5 The applications function

The decision as to whether a specific applications support function is required by the ITO will depend largely on the nature of the

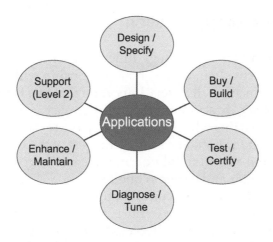

Figure 7.20 IT utility applications functions.

utility services that are to be offered. If the utility service provider is planning to provide 'infrastructure utilities' (backup, storage and so on) then it is unlikely that this function will be required. If, on the other hand, the utility service provider is planning to offer its consumers end user applications (email, HR for example) as utility services, then it will be important to create a center of excellence that is populated by experts in the applications in question. Where this function exists, it is important that close working relationships are forged between it and the engineering and operations functions. This is purely due to the fact that these functions will be providing the basic, but critical, infrastructure services that form the platforms on which applications will rely heavily.

The applications support function can be split into six areas of discipline (See Figure 7.20). In our experience, the traditional ITO tends not to focus very heavily on the various competencies (shown here) that are required to provide and support applications in a very efficient manner. In most cases, the business tends to rely too heavily on the application support contract that exists between the ITO and the software vendor in question. We believe that, in the utility computing model, more emphasis will need to be given to in-house expertise that specializes in application support. This will obviously be particularly relevant where the applications in question are built 'in-house'. In creating this organizational model, therefore, we have gone to some lengths to break out the specific competencies that we believe are required to support an 'application utility'.

7.4.6 The strategy function

We have found that the strategy function is often very difficult to implement effectively in terms of its relationship with the other organizational functions. It is very important that this function is able to create strategic direction for the utility service provider and that it can accomplish this whilst not alienating the rest of the ITO in any way. This function should be seen by the rest of the organization as an 'enabler' and not an 'inhibitor' in the effort to constantly make the IT utility better and easier to operate.

The strategy function is typically populated by thought-leading architects and strategists who (quite naturally) focus heavily on 'the next big thing' in terms of technology or management innovation. There is a part for these individuals to play within the IT utility but, often, a strategy team will create their thoughts and ideas in apparent isolation from the parts of the ITO who are actually running and maintaining an IT service from day to day. If proper attention is not given to the way in which the strategy function works, alongside the engineering and operations functions, for example, then this can lead to a divide being created between the teams. In these instances, the strategy function can find it very difficult to convince the operations groups to take on new ideas (since, often, this will create inherent business risk).

The strategy function can be split into four areas of discipline (See Figure 7.21). In our experience, the most effective method of building

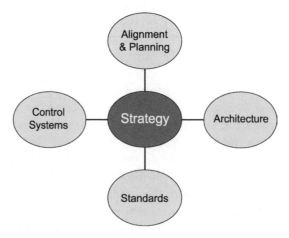

Figure 7.21 IT utility strategy functions.

this competence is to create a 'virtual team', which is comprised of thought-leaders from each of the other centers of excellence. This team may be given management and guidance by a dedicated 'Chief Strategist', who would chair discussions and facilitate the authoring of strategic plans, standards definitions and architecture designs.

7.4.7   The managerial and financial function

The managerial and financial function will, more often than not, actually be fulfilled with the combined competencies of several of the other functions that exist in our model. The competencies that exist here are focused primarily on maintaining senior level relationships with business leaders and in ensuring overall quality (in relation to corporate governance and compliance) throughout the IT utility.

This center of excellence will provide a central point of contact for senior business stakeholders (through its business relationship management discipline) and it will usually form strong links with the 'strategy' function (so as to drive new standards formed through governance initiatives from the 'top down').

As we can see in Figure 7.22, this function can be split into seven disciplines. The skills required to fulfill these are fairly wide ranging and so careful thought and consideration has to be given (whilst planning the IT utility organizational structure) as to how and where these types of skills will be found. This center of excellence is, once

Figure 7.22 IT utility managerial and financial functions.

again, an area of the organizational model that is often best built with outside help (i.e. an expert consulting company) during the initial stages of utility computing transformation.

7.4.8 The asset and configuration function

The asset and configuration function has been modeled with particular focus on industry standards (primarily ITIL) that document 'best practice' with regard to asset change and configuration management technique. These particular functions will often be made the responsibility of an (extended) operations function, and this will require that additional skills and competencies are introduced into this area of the organization. The functions that exist within this center of excellence will become more and more valuable to the utility service provider as the IT utility becomes more complex over time. Without good asset and configuration management discipline, it will prove impossible to measure and analyze certain IT asset attributes that will be very important when running IT as a utility. A good example of this would be the measurement of ongoing IT asset cost, which needs to be understood in some detail in order for the utility service provider to run a successful operation.

As we can see in Figure 7.23, this center of excellence can be split into six disciplines. Project teams should be careful in the selection

Figure 7.23 IT utility asset and configuration functions.

of people that are to fulfill these functions. Our experience is that the IT marketplace is full of individuals who profess to being, for example, 'configuration management experts'. The reality today is that many individuals can show qualifications in the theory of these types of discipline, but very few individuals have proven practical experience of implementing these disciplines in the real world.

8

A Maturity Model for Utility Computing

8.1 OVERVIEW

For many organizations, a utility computing strategy means a gradual move towards the utility model, rather than a radical and instant transformation from traditional IT service delivery to utility computing. In addition, many IT organizations recognize the inherent benefits for utility computing, but do not necessarily feel that they need to deploy all elements of the model in order to reach their desired business goals.

Many of the clients that we have worked with have been interested in exploring the benefits, for example, of virtualization technique (an abstraction of technical complexity so as to simplify IT management and reduce management cost), but do not see the implementation of a 'chargeback' model as appropriate to their business.

For this reason, we felt it necessary to build a model for use as a reference whilst determining the following with regard to developing a new utility computing strategy:

- How far towards a true utility computing model do I need to go?
- At what stage of maturity is my target environment today?
- What benefits will I glean by maturing my IT model?

Delivering Utility Computing.   Guy Bunker and Darren Thomson
© 2006 VERITAS Software Corporation. All rights reserved.

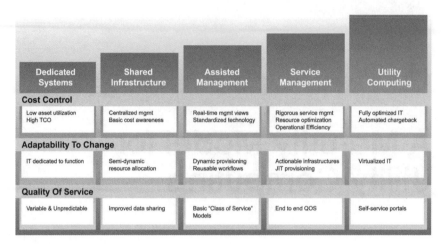

Dedicated Systems	Shared Infrastructure	Assisted Management	Service Management	Utility Computing
Cost Control				
Low asset utilization High TCO	Centralized mgmt Basic cost awareness	Real-time mgmt views Standardized technology	Rigorous service mgmt Resource optimization Operational Efficiency	Fully optimized IT Automated chargeback
Adaptability To Change				
IT dedicated to function	Semi-dynamic resource allocation	Dynamic provisioning Reusable workflows	Actionable infrastructures JIT provisioning	Virtualized IT
Quality Of Service				
Variable & Unpredictable	Improved data sharing	Basic "Class of Service" Models	End to end QOS	Self-service portals

Figure 8.1 The Utility Computing Maturity Model.

As mentioned earlier, targeted business benefits can vary considerably from organization to organization, but the three main areas to which utility computing should bring value are: *cost control*, *agility* and *quality of service*. It was, therefore, important for us to show how these values, specifically, would improve as an IT estate is matured over time.

The Utility Computing Maturity Model allows an IT organization to benchmark current IT estate and plan an IT transformation sensibly and based on real business goals. Figure 8.1 shows the Utility Computing Maturity Model at its highest and most abstract level. Here, we can see basic levels of IT maturity, the fifth being a full deployment and realization of utility computing. The diagram also illustrates some basic attributes of the various maturity levels as they relate to the three main business value areas that are typically focus points for utility computing transformation. This illustration can be used during the very early stages of a utility computing project to illustrate to a senior audience the concept of utility transformation, and to gain input and create debate around how a business's IT estate looks today, in terms of maturity.

It is important to recognize that, in a typical data center, the various areas of IT infrastructure that exist (servers, storage, backup and so on) will almost certainly be representative of different levels of maturity when measured against this model. For example, most IT departments today run their server estates as *dedicated systems*, whereas many organizations have now migrated their storage infrastructure to storage area networking (SAN), and so this type of storage estate would be categorized as *shared infrastructure*.

By targeting a potential candidate for utility computing (see Section 6.3) and then mapping this current estate to the Utility Computing Maturity Model we can start to illustrate and plan how the target will need to be transformed (based on the Utility Computing Reference Model) in order to satisfy a business requirement. More on this later.

8.2 THE MATURITY LEVELS IN DETAIL

First, though, let's explore each of the levels of IT maturity as shown within the model.

8.2.1 Dedicated systems

The dedicated systems model was first seen as distributed information systems evolved out of the shared mainframe infrastructures of the 1970s and 1980s. The comparative reduction in hardware and ownership cost (compared to a mainframe) meant that there was more of an incentive to purchase computing systems (and associated peripherals such as storage) and then dedicate these to a specific purpose. A dedicated system might, for example, be responsible for running an organization's Human Resources application.

Of course, the advantage of this model (and certainly a reason for it still being in use today) is one of asset control. Dedicating compute resources to a specific function allows the 'owners' of that function to be confident that their IT is not at risk of interference from other parts of their organization. In particular, system security often drives a requirement for this model.

High TCO, low utilization

From a financial perspective, new IT project budgets are often used to procure computing systems and it is, therefore, hardly surprising that purchases using these budgets then become 'siloed' and dedicated to the project (maybe a Human Resource system project) that was responsible for purchasing them.

A major disadvantage of this model (in most cases) is one of inappropriate cost. In particular, dedicated systems that are assigned to critical business functions tend to cost far more than they should. This is true because it is normally very difficult to predict how they should be sized. In most IT organizations, standard practice tends to be to over-engineer the system in order to eradicate any future risk associated with system capacity or redundancy. This leads to an enormous amount of wastage and (using the dedicated systems model) there is no way to take back wasted or redundant system resources with a view to reallocating them elsewhere.

Rigid architectures

The rigidity inherent in the dedicated systems model also impacts a business's ability to react quickly to situations. A complaint often levied against this model is that it does not respond well to change. For example, a 'start-up' company may purchase its information systems to run customer transactions, but have absolutely no idea how many customers it is likely to attract during its first year of operations. So, a choice is made; either purchase and deploy a system that is too large but that will not run out of capacity, or risk running out of capacity with an architecture that is difficult to upgrade later (due to dedicated systems needing to be shut down to cater for many administrative tasks such as disk space expansion or memory expansion). In today's competitive business landscape, the ability to provision appropriately expensive IT resources can make all the difference to a business's competitiveness and differentiation (Figure 8.2).

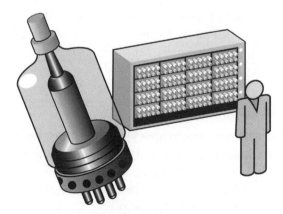

Figure 8.2 Expensive and inflexible – time to think differently.

A 'one-size' QoS model: good or bad?

One could argue that dedicated systems provide the users of IT with an excellent quality of service model (provided, of course, the system is designed appropriately). Certainly, having a system dedicated to a specific purpose allows the designer of the system to focus on the needs of one particular user population. A difficulty in this area, however, can be that a 'class of service' concept can be difficult to implement within a dedicated systems model. To implement 'classes' of the same service with a dedicated system often involves a complex implementation of dedicated system resources for each service class. This will, more often than not, lead to an increase in overall capital asset and operational cost (precisely the opposite effect that a class of service concept should have!).

8.2.2   Shared infrastructure

Shared infrastructures rely on the concept of many applications and users making use of a highly accessible, highly available and highly performing 'shared' resource, normally using a technique of virtualization to simplify the method of access to the asset in question. A good example of this type of infrastructure design today is the storage area network (SAN) as illustrated in Figure 8.3. A SAN allows

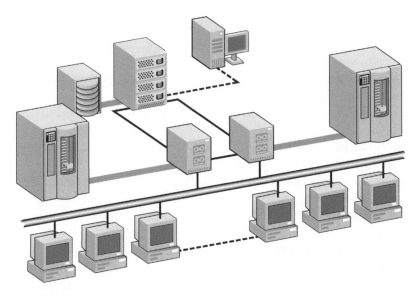

Figure 8.3   A storage area network (SAN).

a business to leverage a significant investment in storage capacity by making this capacity available (across a ubiquitous network) to several applications. Computing systems of varying types can use the same storage, provided they are capable of accessing the storage network and the storage asset itself can be 'fenced' (or *zoned* to use the correct phrasing) so that applications can be secured from one another.

Interestingly, one could consider the traditional mainframe architecture to be a shared infrastructure. Indeed many organizations that have embarked on a utility computing strategy have found that they are, in fact, able to make use of existing mainframe resources (often assets that had depreciated many years ago!) in order to create agile server infrastructures using non-proprietary operating environments, such as Open-Source Linux.

Lowering TCO and improving asset utilization

The shared infrastructure model is adopted typically in order to try to address a few major IT issues. These issues can be summarized as:

- low asset utilization;
- high cost of ownership and management.

By allowing shared access (See Figure 8.4) to infrastructure (such as storage or networking), it is possible (with proper planning) to 'sweat' a technology asset and make it work with a higher utilization level. This is due largely to the ability, in these types of infrastructure, to create 'soft' fencing, meaning hardware assets can be allocated and assigned to applications but then deallocated and reassigned later if this course of action is appropriate. A well-managed shared infrastructure is administered in such a way that, through manual processes, resource can be optimized on a regular basis (typically weekly) in order to keep utilization high (certainly higher than in the dedicated systems model).

Our experience of technology that has matured from a dedicated systems model to a shared infrastructure model, has shown us that asset utilization can be improved by around 30–40% as a result of this effort.

The other inherent benefit, from a cost perspective, of a maturing to a shared infrastructure model lies in the cost of asset management. The theory goes that having the ability to administer centrally, say

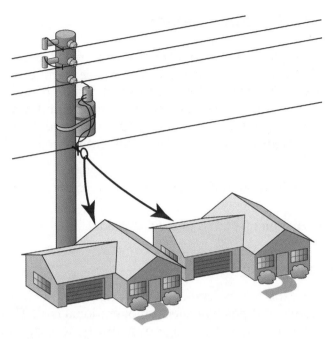

Figure 8.4 Sharing resources reduces unnecessary costs.

a storage asset, should be more cost effective than managing lots of discrete (dedicated) systems. Be careful! Often, your four storage administrators will need to be replaced by one 'super-administrator' in order to cope with the complexity of managing a virtualized, shared storage infrastructure. If your super-administrator costs four times more than the individuals that he replaced, then you have saved nothing!

Semi-dynamic resource allocation (agile infrastructure)

As mentioned earlier, the shared infrastructure model allows administration staff to allocate and deallocate system resources in a more flexible fashion than in the dedicated systems model. Essentially, the process of resource allocation in this model tends to be a manually intensive one. Use of appropriate system diagnostics and monitoring tools tends to be essential if significant efficiency gains are to be achieved through the use of dynamic resource allocation in a shared infrastructure model. The automation of the resource allocation workflows really only starts to occur as we progress to the 'assisted management' maturity level...

8.2.3   Assisted management

In progressing from a shared infrastructure model to the assisted management level of maturity we start to see significant benefit, particularly in the areas of cost control and quality of service (this maturity progression does not tend to have a significant impact on operational agility). In implementing an assisted management model, two major aspects of IT management are matured. These are:

• the standardization and automation of basic operational work-flows;
• the introduction of a 'class of service' model.

Serious cost reduction!

Our project experiences have shown us that these two areas can have a massive and positive effect on operational cost. This is due largely to the ability, at this stage of maturity, to strip out much of the labor cost associated with running IT infrastructure and to replace people who are performing essentially simple, manually (and time) intensive tasks with automated IT workflow engines that can perform these tasks faster, more efficiently and very accurately. In addition, a 'class of service' IT model can, in some cases, allow an IT organization to consolidate aggressively and rationalize its infrastructures so that they are appropriate to the service classes that have been agreed with users. For example, an organization might take its 'shared infrastructure' backup service (with no class of service concept), introduce 'Gold', 'Silver' and 'Bronze' backup service classes and find that 70% of its user population would be satisfied through a subscription to the 'Bronze' service. This should mean that the shared backup infrastructures can be down-sized radically (if, previously, they had all been engineered to provide a 'Gold' service to everybody).

Improved IT/business relations

The introduction of basic automated workflows and a 'class of service' concept can have a very positive effect on the way that the business perceives the services provided by IT. The creation of a 'service catalog' (with associated service level agreements, discussed else

Figure 8.5   Good relationships between IT and the business ensure efficiency and decrease misunderstandings.

where) can help to align IT to the business that it serves. In addition, IT taking on the role of 'service provider' (as opposed to 'technology guardian') can improve drastically the relationship between the IT department and the business lines that it serves (Figure 8.5).

Of the utility computing projects with which we have been involved to date, we have seen the biggest impact on both IT costs and the IT/business relationship occur during a transition from a shared infrastructure model to the assisted management level. It is at this stage of transformation that, politically and culturally, business people start to notice the positive effect that a move towards utility computing is having.

8.2.4   Service management

Once an assisted management stage of maturity has been achieved by an IT department, it is very typical for its inherent benefits (discussed in the last section) to create a willingness and desire to mature IT further to the 'service management' model. This level of maturity largely involves the improvement of the two following areas:

- end-to-end service management disciplines and process;
- extended IT process automation.

In improving and maturing these two areas, an IT department should find significant returns, both in terms of user satisfaction (improved service levels, faster IT support response time and so on) and in terms of cost reduction (largely through further extended automation techniques to complement the newly adopted service management processes).

The evolution to this stage of maturity does not ordinarily see a heavy focus on the technical layers of the Utility Computing Reference Model (i.e. hardware and software), but does involve a lot of change within the organizational and process layers.

ITIL and utility computing

Over the past few years, the ITIL (Information Technology Infrastructure Library) standard has become increasingly popular with organizations that want to focus on improvement of their IT service management processes. The ITIL standard outlines eleven areas of discipline that should be adopted and standardized by an organization that wants to run IT services effectively and efficiently as a business.

The ethos behind the development of ITIL is the recognition that organizations are becoming increasingly dependent on IT in order to satisfy their corporate aims and meet their business needs; this leads to an increased requirement for high-quality IT services.

ITIL provides the foundation for quality IT service management. The widespread adoption of the ITIL guidance has encouraged organizations worldwide, both commercial and non-proprietary, to develop supporting products as part of a shared 'ITIL philosophy'. The eleven ITIL disciplines are broken down between the general disciplines of 'service delivery' and 'service support' and they include:

- Service delivery disciplines:

  - service level management;
  - capacity management;
  - continuity management;
  - availability management;
  - IT financial management.

- Service support disciplines:

  - configuration management;
  - incident management;
  - problem management;

  – change management;
  – service support/helpdesk;
  – release management.

We advise anybody who is serious about maturing an IT estate to the service management level to consider seriously adopting ITIL 'best practice' as part of their transformational program. Actually, much of the philosophy that makes up the Utility Computing Reference Model (described earlier) comes from a synthesis of ITIL and other industry standard best practice methods for creating efficient and reliable IT services.

The project team responsible for IT transformation should not see a standard such as ITIL as something to be either implemented fully or ignored. It is very likely that most organizations will need to adopt only certain ITIL disciplines in order to achieve their project goals. Be careful to scope your IT service management tasks carefully and do not 'over-implement' these standards, this can waste a lot of time and make utility computing projects longer and more complicated than they need to be.

Comprehensive process automation

The IT process automation that addressed basic, simple, manually intensive tasks whilst progressing through the assisted management level of maturity can be extended to good effect once the appropriate service management processes have been modeled and implemented as part of a utility computing project. Many of the workflows that a standard such as ITIL defines can be automated extensively, leading to further cost reduction for an IT organization and fewer administrative or configurative errors (typically caused by human oversight). The requirement and desire further to automate IT processes tends to vary greatly from one organization to another. For many, a fully automated IT estate is both unrealistic and too much of a leap of faith. Careful consideration should be given to how much automation is appropriate within a utility computing transformation.

8.2.5   Utility computing

Progression from the service management maturity level to implementing a true utility computing solution involves focusing largely

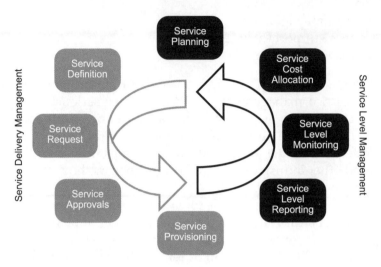

Figure 8.6 A typical utility computing service management lifecycle.

on two areas of transformation:

- IT service financing ('chargeback');
- establishment of a fully virtualized IT service model.

The progression involves maturing the service layer of the Utility Computing Reference Model and ensuring a well-managed service lifecycle (Figure 8.6). Our experience in projects to date has indicated that, in fact, most IT organizations do not feel a desire (or have a requirement) to mature their IT estates to this level of maturity, for reasons that will be discussed later.

An adoption of a true utility computing model will see the relationship between the IT service provider and the IT service consumer (user) change considerably, and this step in a transformation typically has the largest impact on organizational culture.

A true IT utility is seen by its users as a totally 'abstracted' service (where services are requested and provisioned with no thought as to how this has been achieved). Users are bound typically by a very tightly controlled 'class of service' model (discussed later) and have embraced the concept of paying for IT service as part of a chargeback financial management model.

From the perspective of the IT department, most operational procedures are automatic and service management discipline is very tightly controlled (particularly in the areas of change and configuration management). Additionally, the IT department's hardware estate, at this stage, is largely commoditized (i.e. using cheap, redundant and modular components) and managed using virtualized provisioning and allocation techniques (normally driven by heterogeneous software tools).

Chargeback and showback

The very concept of charging business users for IT service (based on capacity usage, transaction volume or something similar) is, typically, 'a bridge too far' for many IT departments today (of course, IT service providers and managed service providers are already very familiar with this model). Our experience has shown us that most IT departments do not want to turn themselves into profitable businesses through the supply of IT service but, rather, want to be able to align their services to specific business requirements in order that the value of IT can be appreciated. The concept of establishing IT as a 'value center' is often a business driver to a utility computing initiative.

Interestingly, the effort involved in creating workflows and reports to provide an insight into IT service cost is required whether an organization wants to do chargeback or not. The concept of 'showback' (generating business reports to show the organization how much IT is costing, but not actually charging anybody for it) is critical to the successful implementation of utility computing. It allows cost transparency to be established and helps the IT department to show its business consumers 'where the money goes'. Having invested in this type of initiative, an organization can switch 'chargeback' on (if required) at a later date with very little additional effort.

A rigid class of service model

Another consideration, when considering a move to true utility computing, is how well the IT consumer will react to a truly abstracted service delivery model, involving an implementation of classes of service.

Utilities are fundamentally restrictive in terms of their method of providing service (See Figure 8.7). This model is certainly not right for every area of the data center. Project teams should analyze their IT service consumers' requirements very carefully before forcing them into a full implementation of IT as a utility.

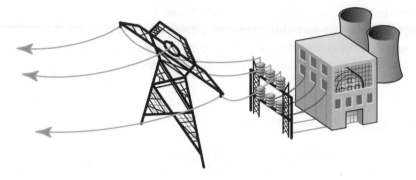

Figure 8.7 A utility at work: a rigid class of service, but provided at lowest cost and maximum efficiency.

8.3 CREATING A UTILITY COMPUTING SCORECARD

Once the utility computing maturity levels are understood, it should be possible to begin the process of mapping a targeted IT infrastructure ('low-hanging fruit') to the maturity model in order to establish a baseline (or assessment of current state). Only then will it be possible to create a plan for appropriate transformation towards the utility computing model. Figure 8.8 shows the five layers of the Utility Computing Reference Model mapped to the maturity model.

Dedicated Systems	Shared Infrastructure	Assisted Management	Service Management	Utility Computing
Service Layer				
No SLA's	Arbitrary SLA's	Basic Class of Service Business Level Reporting	End-End Service Mgmt	Utility Services
Organization Layer				
Distributed Functions Distributed Competence	Centers of Excellence	Simple Service Mgmt Discipline	Comprehensive Svc Mgmt Discipline	IT Value Center
Process Layer				
Bespoke Process	Basic Mgmt Workflows	Routine Task Automation	Comprehensive Automation	Fully Automated IT
Software Layer				
Non-Standardized No Hardware Abstraction	Basic Abstraction Centralized Mgmt Tools	Std Software Tools Basic Auto Provisioning	Service Lifecycle Mgmt Actionable Infrastructure	End-End Utility Mgmt
Hardware Layer				
Distributed, Proprietary	Shared Resources	Shared Asset Pools	Hierarchical Modular Architecture	Commodity Model

Figure 8.8 Detailed maturity model.

Utility Computing Maturity Scorecard

Utility Computing Transformation Target : **Enterprise Backup**

● - Current Maturity ○ - Required Maturity	Dedicated System	Shared Infrastructure	Assisted Management	Service Management	Utility Computing
Service Layer	●				○
Org. Layer	●			○	
Process Layer		●	○		
S/W Layer	●			○	
H/W Layer		●		○	

Figure 8.9 A maturity scorecard for system backup.

Again, we have included in this diagram some indications of the typical attributes that would be expected (within each layer of the reference model) as they correspond to the maturity model.

Appendix C contains a questionnaire, which can be used to assess how a targeted IT estate maps to the Utility Computing Reference Model. Once this type of assessment has taken place, and a baseline has been established, the project team can then begin the process of deciding how each layer of the reference model needs to mature (in the context of agreed business goals).

Figure 8.9 shows how a maturity model 'scorecard' might look once a targeted IT estate has been assessed fully in terms of its current and potential maturity.

Our suggested approach to defining the scorecard is by process of interview (using a questionnaire similar to the one shown in Appendix C) to establish a baseline and then through a process of workshops with project stakeholders to establish and agree appropriate maturity goals (including each layer of the reference model) for the area of IT infrastructure that has been targeted for transformation. Since the five layers of the Utility Computing Reference Model have an impact on many different types of individuals within a business (IT users, infrastructure managers, executives, IT vendors and suppliers, etc.) it will be important to include all appropriate representatives within your utility computing discussions and workshops. Often, these workshops will surface ideas

Figure 8.10   Agree to a plan of action: set appropriate expectations.

and thoughts that had not been considered previously which will ultimately lead to a more robust and compelling business case for transformation.

Capturing estimates of current and required maturity state of an IT estate targeted for utility computing is a critical first step towards the creation of a transformational program plan (Figure 8.10). Our advice is not to continue with a utility transformation until all project stakeholders are comfortable with the output of a maturity assessment. The artefacts that are created as a result of this type of work will, ultimately, become working references that are used to measure project success.

8.4   MOVING UP THE MATURITY MODEL (GENERIC TASKS)

Now let's explore what, in our experience, is typically involved in moving from one maturity level to another. Of course, for the purposes of this book, we have had to keep these descriptions fairly generic but this information should help the reader to begin the process of creating a project plan for a utility computing transformation (Figure 8.11).

Figure 8.11 Executing well-defined tasks reduces unexpected costs.

The following section provides a summary of generic tasks (for each layer of the Utility Computing Reference Model) that are typically required in order to mature an IT estate towards the utility computing model.

8.4.1 From dedicated systems to shared infrastructure

- **Tasks within the service reference model layer:**

  1. Run 'service establishment' workshops to establish OLAs (Operational Level Agreements).
  2. Provide routine OLA reports.

- **Tasks within the process reference model layer:**

  1. Capture all routine tasks as 'working practice'.
  2. Establish standard (manual) workflows for service support.
  3. Establish management controls for workflow tracking.

- **Tasks within the organizational reference model layer.**

  1. Create a Service Level Manager role for business liaison.
  2. Create IT management centers of excellence.
  3. Conduct IT staff skills analysis and train appropriately.

- **Tasks within the software reference model layer:**

  1. Create a basic virtualization layer to abstract hardware complexity.

2. Implement system monitoring to track OLAs.

3. Create a software standards 'blueprint' for ongoing design.

4. Implement agents to monitor performance and availability.

- **Tasks within the hardware reference model layer:**

1. Build an accurate hardware asset/configuration management database.

2. Monitor hardware resource utilization for capacity planning.

3. Create a hardware standards 'blueprint' for ongoing design.

8.4.2 From shared infrastructure to assisted management

- **Tasks within the service reference model layer:**

1. Extend OLAs to provide routine SLA reports with trend data.

2. Create a service catalog for IT infrastructure services.

3. Create basic marketing and awareness for catalogued services.

- **Tasks within the process reference model layer:**

1. Define integration between operational and management workflows.

2. Create automated workflows for key operational activities.

3. Evolve service support disciplines to industry best practice.

- **Tasks within the organizational reference model layer:**

1. Extend SLA Manager role to encompass new service catalog.

2. Mature centers of excellence to map to service support best practice.

- **Tasks within the software reference model layer:**

1. Implement monitoring to track SLAs.

2. Design and implement IT 'views' of infrastructure, performance and availability.

3. Implement basic IT management 'framework'.

4. Create a software standards 'blueprint' to map directly to catalogued services.

- **Tasks within the hardware reference model layer:**

  1. Commoditize storage hardware pools, where appropriate.
  2. Implement tiered hardware model.
  3. Conduct hardware consolidation exercise.

8.4.3 From assisted management to service management

- **Tasks within the service reference model layer:**

  1. Create comprehensive SLA reporting for business executives.
  2. Implement a formal service request handling process.
  3. Implement formal IT helpdesk facilities.

- **Tasks within the process reference model layer:**

  1. Implement all appropriate service delivery and service support best practice (ITIL or similar).
  2. Automate workflows for availability, performance and system management.
  3. Implement management 'portal' for all IT service reporting.
  4. Define classes of service for infrastructure services.

- **Tasks within the organizational reference model layer:**

  1. Ensure that the centers of excellence model is extended to encompass all service delivery/support disciplines.
  2. Recruit dedicated, specialist staff to support service management processes (configuration management experts, etc.).

- **Tasks within the software reference model layer:**

  1. Implement correlation software to support IT management.
  2. Extend virtualization layer to support classes of service.
  3. Implement standard workflow tool for operational processes.

- **Tasks within the hardware reference model layer:**

  1. Implement shared server model.
  2. Commoditize/consolidate service estate, where appropriate.

8.4.4 From service management to utility computing

- **Tasks within the service reference model layer:**

  1. Implement consumer 'self-service' portal.
  2. Implement service consumption measurement.
  3. Operate CRM/PRM functions (see Utility Computing Reference Model).
  4. 'Run IT as a business' (profit center).

- **Tasks within the process reference model layer:**

  1. Extend workflow automation to service delivery/service support best practice and service lifecycle support.
  2. Implement rigorous process compliance auditing.
  3. Integrate operational and management workflows.
  4. Define 'continuous improvement' procedure for processes.

- **Tasks within the organizational reference model layer:**

  1. Recruit staff to manage CRM/PRM functions.
  2. Conduct IT staff audit and skills review and reorganize as appropriate.
  3. Ensure that all functions of the Utility Computing Reference Model have skills mapped to them at an appropriate level and standard.

- **Tasks within the software reference model layer:**

  1. Implement a 'Service Governor' to automate dynamic allocation of IT resources.
  2. Implement 'dynamic application workload management'.

- **Tasks within the hardware reference model layer:**

  1. Extend commoditized, modular hardware model.
  2. Create virtualized server and storage resource 'pools'.

9

A Transformational
Approach

9.1 OVERVIEW

Having embarked on several utility computing transformations
with large enterprises in recent years, it quickly became apparent
to us that, in fact, all transformations, regardless of how radically
they will transform an IT infrastructure (i.e. how far up the maturity model they will take an IT estate), should follow the same basic
model for transformation.

The Utility Computing Transformational Model that we have developed is aligned with a program lifecycle that an organization
might reasonably adopt when planning a major initiative. It is comprised of four distinct phases:

1. Prepare.
2. Model.
3. Transform.
4. Operate/innovate.

The lifecycle alignment of the Utility Computing Transformational Model is depicted in Figure 9.1.

Delivering Utility Computing.   Guy Bunker and Darren Thomson
© 2006 VERITAS Software Corporation. All rights reserved.

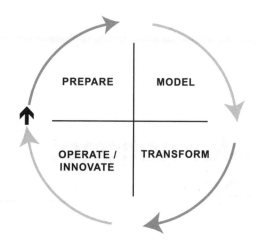

Figure 9.1 Transformational methodology.

9.1.1 Summary of transformational phases

Let's explore briefly what each of the phases of the model should strive to achieve before looking at the transformational model in more detail.

Prepare phase

In the prepare phase, the current IT function is assessed from a number of perspectives (not just technology) and the facilities that could benefit from being (re) configured using the utility paradigm are identified.

Model phase

During the model phase, a technology architecture and management framework for the utility is developed, and the services/ service classes that will be made available defined. The total cost of ownership for the current state (non-utility) and the future state (utility) are modeled in preparation for the creation of a business case.

Transform phase

The transform phase is comprised of two projects – one to implement the appropriate management and control functions and policies, and the other to (re) configure the technology assets to create a utility service.

Operate/innovate phase

In the operate/innovate phase, the appropriate facilities for managing consumers and providers of the service, and for maintaining the service portfolio, are introduced. Additionally, a utility assessment (review) is conducted, to assure ongoing evolution and continuous improvement of the IT utility.

9.1.2 The value of the Transformational Model

To implement a utility computing service, a customer typically will follow the four phases of this model in a cyclical manner, commencing with the prepare phase and concluding with the operate/innovate phase.

Although each phase may be considered as an independent 'project' (i.e. largely self-contained with stated outcomes and deliverables), certain interdependencies exist between the phases, in that the outcome/deliverables from one phase provide the input necessary for a subsequent phase.

It is recognized that not all customers will want, or be able, to implement fully a utility computing maturity level; however the IT Transformational Model is designed to achieve real value for all customers through clearly defined activities, outcomes and deliverables.

The associated value that may accrue from undertaking such a project could include:

- increased knowledge and awareness of the benefits of adopting utility computing principles;
- plans for organizational realignment or improving operational effectiveness;
- quantification of cost savings that could arise from standardization and virtualization;
- heightened awareness of techniques to improve the relationship between IT and the business users.

9.1.3 The Transformational Model in detail

Figure 9.2 gives a logical representation of the Utility Computing Transformational Model in detail.

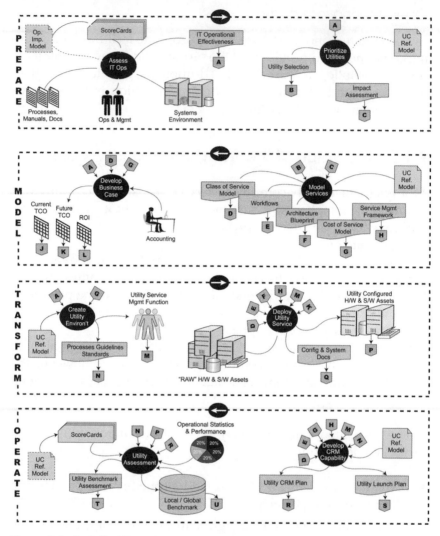

Figure 9.2 Detailed Transformational Model.

9.1.4 A close look at the Transformational Model phases

Within the Transformational Model we have defined two distinct work units (outputs) to support each of the four phases. These work units, shown as circular objects in the Figure 9.2, are:

- In the prepare phase:

- an IT operational assessment;
- utility selection and impact assessment.

- In the *model* phase:

  - utility service modeling;
  - business case preparation.

- In the *transform* phase:

  - utility environment preparation;
  - utility service deployment.

- In the *operate/innovate* phase:

  - utility launch and consumer relationship management;
  - utility operational assessment.

Each of the work units within the Utility Computing Transformational Model is designed to recognize and address these challenges and provide tangible deliverables – these are denoted [X] in the paragraphs below to correspond with the diagram labels shown in Figure 9.2.

IT operational assessment

The initial work unit should calibrate the effectiveness of the existing IT operation against industry-standard best practice metrics and the Utility Computing Reference Model, to provide a measurement baseline – [A], the 'Operational effectiveness report'.

Utility selection and impact assessment

This work unit combines utility service definition and a prioritization with associated parameters – [B], the 'Utility selection report', with an assessment of the type and impact of the changes (and associated risk) that will result from implementing a utility model – [C] the 'Utility impact assessment report'.

Utility service modeling

The scope of this work unit includes the development of a services model – [D], associated operational workflows – [E], an architectural blueprint – [F], an indicative cost model – [G] and a management framework specification – [H].

Business case preparation

This financial modeling work unit includes the construction of TCO models for current-costs – [J] and future costs – [K] that support a financially robust ROI calculation – [L]. These TCO models include ongoing operational costs.

Utility environment preparation

This project engagement handles the organizational change aspects of 'people and process' that are required to support the operation and management of the utility model resulting in a utility service management function – [M] and associated management processes documentation – [N].

Utility service deployment

This work unit handles technology implementation, configuration and testing to support the defined classes of service and results in a utility infrastructure – [P] and associated documentation [Q].

Utility launch and CRM

This work unit develops a plan for consumer relationship management – [R] covering both service consumers and service providers (which will be new to many IT organizations) and a utility launch plan – [S].

Utility operational assessment

This work unit provides an objective calibration of the effectiveness of the utility model, documented in the form of a Utility benchmark assessment report – [T]. The associated benchmark results would normally be captured in a database – [U] and compared with historic data to show trends and identify key performance areas to target for continuous improvement.

Figure 9.3   Proper preparation prevents poor performance.

9.2   THE PREPARE PHASE

The prepare phase considers the current configuration, effectiveness and capability of an IT function from the perspective of its readiness to adopt the principles of utility computing, and evaluates the activities, impact and risks associated with the introduction of a utility computing service (Figure 9.3).

At the conclusion of this phase, the program sponsor will be informed fully of the nature and scope of a program of activities required to adopt a utility computing approach and implement a utility computing service.

As discussed earlier, this phase is comprised of two activities:

- operational assessment;
- utility selection and impact assessment.

9.2.1   The operational assessment

The purpose of the operational assessment work unit is to calibrate the current status of the IT function in terms of the effectiveness of its:

- organization;
- operation;
- process;

- technology;
- reporting.

Current state data is gathered, and the calibration is established, using scorecards developed from IT industry best practice models. Our experience has shown that the following industry standard models should probably be considered for scorecard development:

- Information Technology Infrastructure Library (ITIL);
- Control Objectives for Information Technology (CoBIT);
- Capability Maturity Model (CMM).

This data is supplemented with observations made during the assessment and with information disclosed during interviews with IT and business personnel. The data, observations and information are analyzed and assessed in order to:

- establish current levels of resource (hardware, software and human) effectiveness;
- determine current maturity of operations and management processes;
- assess the level of alignment between IT and the business.

The assessment is documented in the form of an IT operational effectiveness report and presented to the program sponsor.

9.2.2　Utility selection and impact assessment

The purpose of the utility selection and impact assessment work unit is to establish the impact to the existing IT function of adopting the principles of utility computing and to determine a candidate service and an outline scope for an initial utility computing service.

A comparison is made between the current state/capability of the IT function, as defined in the IT operational effectiveness report, against the specifications provided by the Utility Computing Reference Model (Figure 9.4). This comparison enables a gap analysis and impact assessment to be produced, covering the following elements of the IT function:

- organization;
- operation;

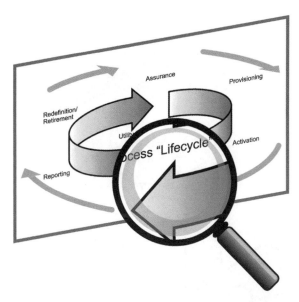

Figure 9.4 Without continuous assessment, progress is hard to measure and success even harder.

- process;
- technology;
- reporting.

The outcome of the comparison is documented in the form of a utility impact assessment report that:

- details the nature and scope of the changes required to introduce a utility computing service;
- quantifies the risks associated with introducing and not introducing a utility computing service.

The work unit should also evaluate and prioritize candidate infrastructure services for implementing as a utility computing service. These, in our experience, might include:

- backup;
- disaster recovery;
- storage management;
- server management;
- performance of application-enabling infrastructure.

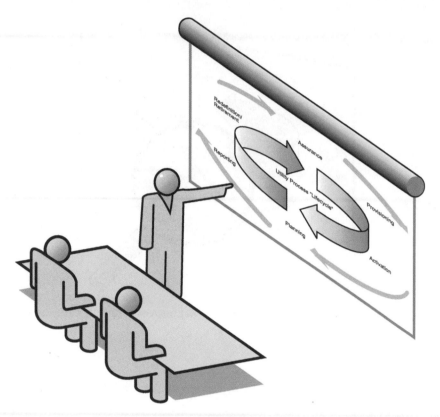

Figure 9.5 Executive buy-in at board level is critical for success.

This evaluation should be documented in a utility selection report that contains a definition of the chosen utility computing service, together with a high-level activity roadmap with key milestones and deliverables for the introduction of that service. The utility impact assessment and the utility selection reports are delivered and presented to the program sponsor (Figure 9.5).

9.3 THE MODEL PHASE

The model phase considers the introduction of a utility computing service from the perspectives of consumer and provider, and defines the classes of service to be offered, together with the associated indicative costs. It addresses the financial impact of introducing a utility computing service by modeling the current costs

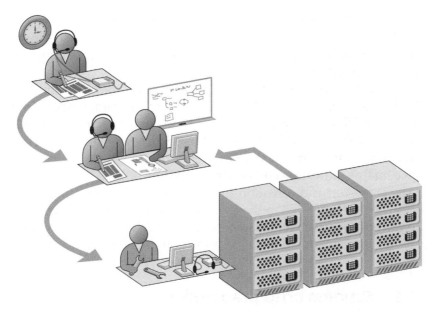

Figure 9.6   Model the effect; even a crude model can pay huge dividends.

and likely future costs, and develops TCO and ROI scenarios to underpin a business case for an adoption/implementation program (Figure 9.6).

At the conclusion of this phase, the program sponsor will be fully aware of the scope and costs of introducing a utility computing service.

This phase is comprised of two work units:

- utility service modeling;
- business case preparation.

9.3.1   Utility service modeling

The purpose of the utility services modeling work unit is to determine the overall service model for the selected utility computing service (as defined in the Utility selection report) and to create the associated class of service definitions.

An iterative consultative process involving business and IT management determines the business requirements that define the overall scope of the services model and the range of service classes required to support it. This process is directed by applying the specifications detailed in the Utility Computing Reference Model and aligned with the phases defined in the Utility Process Lifecycle Model (see Section 7.3).

The output from this process is incorporated into a Utility service architecture blueprint that forms a reference framework for the service model by defining standards for component selection (hardware and software) and service characteristics (availability, performance and cost parameters) for each class of service. This document also contains indicative costs for each class of service.

The provision and consumption aspects of the utility computing service are documented in the form of a utility service management framework, which covers the operational, management and reporting elements associated with the provision of the utility service.

9.3.2   Business case preparation

The purpose of the business case preparation work unit is to establish the impact of implementing a utility computing service in financial terms, and to refine the indicative costs for the various classes of service defined in the Utility service architecture blueprint.

A consultative process would normally be undertaken to establish the current cost base for providing the same facilities as will be made available through the utility computing service. In the absence of customer-specific current cost data, calculations are based on commonly accepted industry figures.

Elements of the current cost base should include:

- hardware procurement;
- commissioning;
- environment;
- depreciation;
- utilization rates;
- maintenance and administration;
- monitoring and operations;
- decommissioning.

The cost base forms the basis of a current state total cost of ownership model (TCO).

A comparison model should then be defined for the provision of the same facilities in the form of a utility computing service based on the Utility impact assessment report, and a comparison of the two models undertaken. This financial modeling exercise creates an information set that supports:

- cost-of-service modeling;
- sensitivity analysis;
- ROI modeling.

The output from this consultative process enables usage costs to be established for each of the defined service classes, according to the organization's desired business model (make profit, break even, or other).The output is documented in the form of:

- a current state cost model (TCO);
- a utility service cost model (TCO);
- a utility service assumptions and cost-of-service model;
- a utility service ROI model.

These documents facilitate the construction of a financially valid business case to support the implementation of the utility computing service.

9.4   THE TRANSFORM PHASE

The transform phase readies the IT and business functions for the introduction and operation of a utility computing service and then manages the deployment of the initial service.

These are transformational projects, which are almost certain to impact the existing IT and business functions.

Due to the nature of the activities undertaken in this phase, considerable cross-functional collaboration is required and the commitment/engagement of senior management in both IT and the business is a prerequisite for a successful outcome.

At the conclusion of this phase (Figure 9.7), a *utility service management* function will have been created with responsibility for managing all aspects of consumer and provider relationships, and portfolio management for utility services. An initial utility computing service will have been implemented and tested, ready for launch.

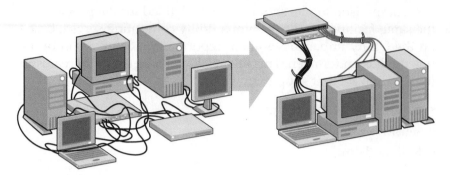

Figure 9.7 The transformation.

This phase is comprised of two activities:

* utility environment preparation;
* utility deployment.

9.4.1   Utility environment preparation

The purpose of the utility environment preparation work unit is to establish the appropriate organizational structure and institute the appropriate processes to support the operation of a utility computing service, with particular emphasis on portfolio management.

This is a transformational program that relies heavily on the Utility Computing Reference Model and impacts many areas of IT, including:

* organization;
* operation;
* process;
* reporting.

The constructs, guidelines and principles contained in the Utility Computing Reference Model are adapted to create the utility service management function defined by the model, taking into account customer-specific requirements as defined in the utility service management framework that has been defined in the modeling phase.

The utility service management function will be responsible for all aspects of service management, including:

- consumer relationships;
- provider relationships;
- portfolio management.

This function operates through a suite of processes, which are documented in the process layer of the Utility Computing Reference Model.

Depending upon the size and maturity of the existing IT function and the scope and nature of the initial utility computing service, the utility service management function may be implemented as:

- a separate organizational entity; or
- an overlay (or matrix) function to the existing IT organization; or
- a 'virtual entity' (i.e. one or more person(s) within the existing IT function to whom responsibility for this function is assigned).

The utility environment preparation activity is completed by defining initial targets for its key performance areas and documenting its organization, structure, terms of reference and processes in a *Utility Service Management Function – Organization and Operations* document.

9.4.2 Utility service deployment

The utility service deployment activity defines and executes a cross-functional project to establish the initial utility computing service in accordance with the design criteria from utility services modeling.

During this phase, the utility service management function, and selected resources from within the existing IT function, implement and configure the appropriate infrastructure environment to support the utility computing service. Each class of service will be implemented by referencing the Utility service architecture blueprint and by following a predefined service lifecycle, ideally incorporating the following phases:

- planning;
- provisioning;

- activation;
- assurance;
- consumption;
- redefinition;
- retirement.

On completion of implementation activities, each service class will be tested and validated, together with the appropriate monitoring and management facilities.

Service performance will be verified against the defined performance metrics and baseline data captured for performance and capacity modeling to ensure the service will meet, or exceed, its design objectives.

Upon completion of this phase, a validated utility computing service will be available and ready for launch. The service will be complete with the appropriate consumer interfaces, management facilities and operational processes.

9.5   THE OPERATE/INNOVATE PHASE

The operate/innovate phase addresses the ongoing operational management requirements for the utility computing service and defines an inspection process for monitoring and assessing the performance of the service in order to drive continuous improvement and further innovation.

The operation of a utility service calls for particular activities to be undertaken in support of the consumers and providers of the service, and to manage the portfolio of available services. The necessary organizational constructs will have been formalized during the utility environment preparation activities and these will now be activated in accordance with the Utility Computing Reference Model.

The inspection process considers the operation of the utility computing service from the perspectives of consumer and provider, and calibrates the performance against defined key performance metrics for the class of service.

This phase is comprised of two activities:

- utility launch and CRM;
- utility operational assessment.

9.5.1   Utility launch and CRM

The purpose of the utility launch and CRM work unit is to define and establish the appropriate mechanisms within the utility services management function to launch the service effectively and manage:

- the consumer relationship;
- the provider relationship;
- the service portfolio.

Service portfolio management is addressed largely within the utility environment preparation work unit. The utility launch and CRM activity is focused on consumer relationship management and provider relationship management. Through a consultative process, the specifications contained in the Utility Computing Reference Model are adapted to suit the customer's specific utility service, organization structure and management requirements, as documented in the Utility Service Management function. The scope of the defined processes and operational best practice guidelines would normally include:

- sales and marketing;
- service delivery assurance and support;
- communications and notifications;
- service performance and usage reporting;
- billing and account management.

The output from this process is a *launch program* detailing the activities associated with the utility computing service inauguration and a defined and documented suite of operational processes and guidelines for the effective management of the utility computing service.

9.5.2   Utility operational assessment

The utility operational assessment activity is designed to calibrate the current status of the utility computing service in terms of the effectiveness of its:

- organization;
- operation;

- processes;
- performance;
- relationships.

Current state data should be gathered using scorecards developed from IT industry best practice models, including:

- Information Technology Infrastructure Library (ITIL);
- Control Objectives for Information Technology (CoBIT);
- Capability Maturity Model (CMM);
- Key performance targets for each class of service.

This data would then be supplemented with observations made during the assessment and with information disclosed during interviews with service providers and service consumers.

The data, observations and information are analyzed and assessed in order to:

- establish current levels of operational performance;
- determine current maturity of operations and management processes;
- assess the quality of the relationships with consumers and providers;
- create current baselines for the above.

Figure 9.8 Without innovation the ITO will become obsolete.

The assessment is documented in the form of a Utility Operations Report and compared with the previous baseline (if available) to:

- analyze performance and effectiveness trends;
- identify areas of slippage for correction;
- quantify cost-saving predictions used in the business case;
- identify target areas for further improvement.

The Utility Operations Report and Comparison document facilitate a process of continuous improvement in the management of the utility computing service (Figure 9.8).

10

Technology for Utility Computing

10.1 OVERVIEW

In Chapter 3 we looked at the historical trends that have enabled the utility computing revolution. Trends such as the increase in bandwidth and the rise of the Internet are only part of the story; the most significant recent development is the progress of software to support the new utility computing paradigm. This chapter looks at the technology needed for both infrastructure and service management and gives guidelines to help in successful selection of appropriate tools (Figure 10.1).

Successful technology selection and deployment enables the IT organization (ITO) to actually develop and deploy IT as a service, in a true utility manner. It is software that is driving the utility computing revolution and, while monitoring, reporting and automation are all important, the revolution started with virtualization.

10.2 VIRTUALIZATION

If there is one word that has become the IT buzzword for the start of the new millennium it is *virtualization*. Unfortunately, it has been over-hyped and much of the true value lost in the multitude of different solutions put forward by different vendors.

Delivering Utility Computing. Guy Bunker and Darren Thomson

IT Strategy

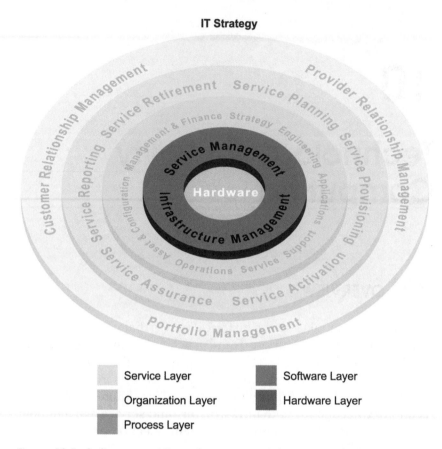

Service Layer

Organization Layer

Process Layer

Software Layer

Hardware Layer

Figure 10.1 Software and the reference model.

In essence, virtualization is the introduction of a technology layer that abstracts specific features, usually hardware, such that the application running on top does not need to understand the hardware below, to the point where it can be changed without the application realizing. This means that applications can be moved from one server to another to improve performance or increase efficiency within the environment. Storage can also be virtualized, enabling equally efficient use of resources, as can the network. Virtualization is probably the most important component in the value proposition for the ITO when considering the move to utility computing, as it enables applications to be sized for their requirements today, rather than having to be sized for what they might need in two or three years' time; it therefore enables the ITO to make more use of its budget.

We shall examine some general guidelines when looking for virtualization software, and then look at some of the details for each of the following:

- storage;
- server and application;
- network.

10.2.1 General guidelines for buying virtualization tools

The data center and IT environment of today is not homogeneous. While an obvious statement, it is important to remember that in moving towards a utility environment and delivering IT as a service, one of the biggest gains is in standardizing the various approaches and processes. Selecting virtualization tools that help manage the heterogeneous environment, rather than ones that just manage a single platform or device, will drive the standardization.

When purchasing software tools for virtualization, examine all the ways in which you might use them. For example, will they be useful for a consolidation program? Can they help with data and application migration? Will they enable a flexible environment where applications and data can be moved so they best fit the current resources? When implementing service levels, will the tools help with changes in service levels?

Centralized control and reporting are important. Can the tools report on all the resources in the environment, can they manage them remotely as well? A consistent policy for resource management tools, combined with a well-documented architecture, will drive the structure of the environment and the ITO towards that needed for utility computing.

10.2.2 Storage

Storage virtualization has been around for decades, and within open systems for more than fifteen years. As with all virtualization the goal is simple, abstract the hardware away from the application so that it can be changed without the application having to be stopped and restarted.

Storage virtualization has become an overused term and now every storage vendor touts their version of virtualization as being the best. In-band, out-of-band, block, disk, file system, host-based, network-based, storage-subsystem-based, they are all talked about and often in the same breath. Unfortunately, there is no right answer when looking for a storage virtualization solution and so it depends on what it is required to do. The topic is worthy of a book in its own right and indeed several have been written (for example Massiglia and Bunn, 2003).

RAID (redundant array of inexpensive disks) is probably the most common storage virtualization used. Either it is carried out in the hardware of the array or it is done in software; either way it ensures the availability of the data should a disk fail. With the advent of the storage area network (SAN), storage virtualization became significantly more attractive. By increasing the connectivity using fabric switches, SAN-wide aggregation using virtualization software, such as VERITAS's Storage Foundation^{TM} Suite or EMC's Storage Manager, to create storage pools became a reality. Pooling storage and being able to allocate it flexibly to multiple servers is extremely valuable outside of utility computing, but within utility computing, this flexibility enables improved resource utilization by allocating only what is needed. Other features, such as being able to grow the size of the file system when it needs more space, mean that applications can be deployed with only the storage they require at that instant in time, rather than having to size for potential peak requirements.

Storage virtualization software is becoming more sophisticated and the ability for multiple tiers of storage to be used within a single file system is now possible. Storage properties can be defined and performance information used to enable the system to store the most frequently used data, or the most important data, on one tier of storage, while the rest is on other tiers. Valuable resources can be freed up by moving old, or less frequently used, data automatically from one tier to another.

With the increasing complexity of the storage environment, it has become essential that the system aid the administrator with provisioning and allocation. IT workflows can be used to assist with provisioning. Similarly storage allocation tasks can be fully automated based on policy. Both of these can be considered as options as they will help move the onus from the system administrator down to the operator when dealing with requests. In most ITOs it is still a requirement for insight, if not full control, for all layers, from the logical to the physical, and so tools should be chosen that enable this.

Virtualization tools that enable data to be moved from one array to another should be considered important for future-proofing and, while not completely applicable yet, migration from one OS to another might also be considered.[1]

In this ever-changing world, perhaps the most important aspect of selecting a storage virtualization tool is to look for features which will insulate the ITO from requiring new management skills in order to deploy new storage technology. For example, when SANs were first introduced, these managed the storage with the same tools that were already being used to manage direct attached storage (DAS). In the future, when iSCSI is deployed, the same tools should be able to be used. Storage is one area where centralized control is important from an efficiency and effectiveness perspective. Selecting a tool or tools that give a view on all storage and enable cross business reporting is important.

Key questions

- Can the management application run on multiple platforms? Can it be used to manage a heterogeneous OS environment?
- Is the virtualization consistent across multiple OS platforms?
- Does the application support all the storage vendors you require?
- Does the application support all the storage architectures you require?
- Will the application insulate you (as far as possible) from future technology advances?

10.2.3 Servers

If virtualizing storage is one half of a utility computing infrastructure environment, then efficient management of servers is the other. Servers have become commodity items, with CPU performance

---

[1] This is not really applicable yet as most applications are unable to read their data in an OS-independent manner. However, this is beginning to change with applications writing data in byte-independent formats so that they can be transferred. Linux appears to be the driving factor for this.

making even 'cheap' solutions capable of delivering business critical services. Servers are no longer booted in isolation, technology to boot them across the network, or even across the SAN, has given the ITO greater flexibility and this ease of configuration, coupled with commoditization, leads to virtualization. Virtualization, when used in conjunction with a 'server', can be interpreted in a number of different ways:

1. Server provisioning/reprovisioning.
2. Application migration across servers.
3. Application migration using virtual machines.

For the ITO, server provisioning is something that is done already on a regular basis and often standard builds are employed to make this as simple as possible. However, when looking to reprovision on demand, a new set of requirements is introduced to manage the server lifecycle (See Figure 10.2).

New architectures based on blade servers have been one of the driving forces behind the change. Based on x86 chips and capable of running different network operating systems (NOSs), blade racks

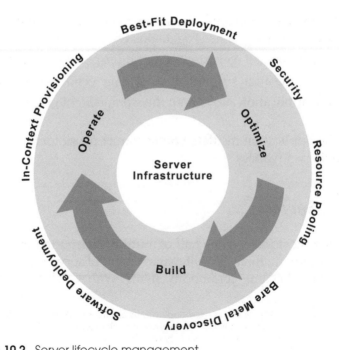

Figure 10.2 Server lifecycle management.

are becoming a familiar sight in data centers. Currently used for the web tier and as edge servers for applications, their performance has increased such that they are creeping into the realm of the application and database server.

New versions of server lifecycle management software, such as VERITAS's OpForce™ or IBM's Tivoli Intelligent Orchestrator, now handle more than just installing and maintaining the OS. Applications can be installed and customized based on deployment templates, and storage allocated and reallocated as required.

Just being able to build and rebuild systems efficiently and consistently helps drive down costs in a utility environment. Utilization of the resource is increased by being able to repurpose rapidly based on need.

Note for system administrators

When reprovisioning a server, it is worth considering the data that used to be held on it. Does all data need to be deleted, including things like configuration files? If so, then is a simple disk format enough, or should a more stringent method, such as the US Government's Department of Defense standard for overwriting all data, be used?

Often, when people talk of server virtualization, they are thinking if it from the application's perspective. Is it possible to create an environment where an application can run on any one of a number of servers, and not only will the application not care which server, but neither will the consumers of the application? Clustering applications have been around for several decades but it is only in the last five or six years that they have become commonplace in open systems. Initially created as a more efficient high availability/disaster recovery solution than requiring duplicate 'hot standby' systems, they are now being used more often as a means for load balancing. Clustering software has a much finer granularity than the 'all-or-nothing' approach of high-availability hot spares, enabling individual applications to be distributed around the cluster as required (Figure 10.3).

Clustering solutions today tend to manage homogeneous OS server clusters with relatively small numbers, for example 16 or 32 nodes. Well defined configurations can be created and because the application is known by an IP address, rather than a server address,

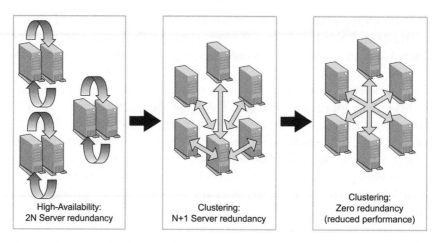

Figure 10.3 From high availability to clustering servers.

it can be moved around the cluster as needed and based on load. Applications can be deployed initially on smaller, low power servers and then moved to larger ones as required.

For the ITO in a service driven organization, such flexibility is exceedingly important. Depending on how SLAs are drawn up, customers can have their application running on one class of server (with the associated cost) and then change to another on demand (i.e. they change levels of service). Alternatively, the ITO can move applications around to ensure that performance based SLAs are met and to free up resources if necessary.

Next generation clustering and workload balancing solutions not only improve scalability up to several hundred servers, they also are introducing technology that allows applications to be moved without loss of connection state. What this means is that even partially complete transactions can be moved seamlessly, with the consumer only noticing a slight glitch in performance when the application actually moves.

The final piece of technology that is emerging as part of the server virtualization landscape is the *virtual machine* (VM).[2] Nearly all major OS vendors are either shipping, or have announced that they will ship, server virtualization software as part of their standard OS (mostly notably Sun with their release of Solaris version 10). In addition, a number of other vendors also ship solutions, including

---

[2] Not to be confused with volume manager, which is also abbreviated to VM.

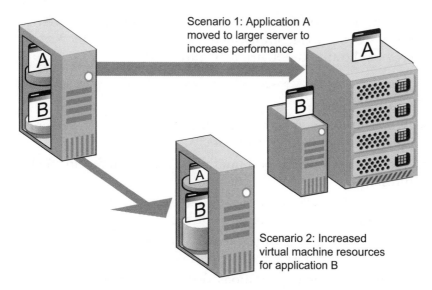

Scenario 1: Application A
moved to larger server to
increase performance

Scenario 2: Increased
virtual machine resources
for application B

Figure 10.4 Using virtual machines for server and application mobility.

EMC with VMWare and Microsoft with Virtual Server. VMs are useful in a number of ways, with consolidation being the most popular. When looking at consolidation opportunities, it has become apparent to many businesses that cohosting applications on the same server in the same OS is too risky,[3] especially when having to deal with patch levels and drivers that end up being specific for the application. (Not that other patch levels will not work, more that the testing has not been done and so people are unwilling to risk it.)

VMs are also being used as a mechanism for load balancing and limiting resource usage, both of which play into the utility computing environment. With applications existing in their own VM, functionality exists to hibernate the running VM, along with the applications it contains, and to migrate it to a different server and restart it (Figure 10.4). While not as granular as the technology that migrates individual applications, it is gaining in popularity and moving from being a development tool to one which is used in production environments. Improved management by the server machines means that VMs can have their resources throttled on demand. Again, this control will enable the ITO to build services around VMs and then to offer service levels based on how much resource is required. By

---

[3] Risky from a business perspective, where application isolation is needed so that if one application were to crash, it would not bring down the entire server.

basing it on a VM, if the service level needs to be improved, additional resources can be allocated instantaneously to the VM and, therefore, the application inside (Figure 10.4).

Key questions

- What is it that you want to do? Server provisioning, server and application configuration or both?
- Does the software support multiple server OS and chip architectures?
- Does the software support multiple architectures, for example, standalone servers as well as blade racks?
- What level of granularity is required for virtualization? Is it the whole server, or just a single application?
- When transferring applications, does state need to be preserved?
- Can the software enable application transfer between dissimilar hardware?

10.2.4 Network

The term network is now confusing, while most people immediately think of IP-based networks (the Ethernet), within the ITO it is also synonymous with the SAN. The fundamental difference being that the fiber channel network of the SAN is usually managed by the ITO, whereas the IP network has always been off-limits and managed by a different group. Often seen as a *given*, it is always there and when it is not, it is a disaster, as everything stops. The rise of iSCSI as a technology means that there is now a crossover, iSCSI storage networks are directly in the path of the standard IP network, and while in many cases they are run using completely separate HBAs and wiring, the tools to manage them are often exactly the same. The ITO and the network groups will have to work more closely in the future to resolve issues.

Virtual local area networks (VLANs) are increasing in popularity, especially when it comes to architectures to support utility computing. In the new architecture, there is a great deal of flexibility offered, and with virtualization technology such that applications can be moved seamlessly from one server to another, this introduces issues regarding scalability and security, both of which can be resolved using VLANs. By introducing VLANs and managing them when provisioning servers, it is possible to group servers together for specific tasks. When the task has finished, or the server is required for a different one, the VLAN can be changed to put it logically into a different group, which may have different users and security demands.

Key questions

- Do you have a separate group that runs the network? Do they work with the ITO?
- Are you considering iSCSI?
- Are VLANs a potential solution for you to use to segregate your network?
- Is VLAN management to be integrated into your provisioning product?

10.3   PERFORMANCE MONITORING

Performance monitoring tools are essential within the utility computing environment, not just for spotting performance issues, but also for looking at trends in order to resolve anomalies before they become problems. There are two ways to look at performance. The first, and most important, is from the perspective of the user. The tools selected for performance monitoring need to be able to monitor from the user through the various middle tiers to the back-end disk storage, i.e. from end to end. Data needs to be collected from each tier and then correlated across them. Trending and reporting are essential in order to spot where degradation in performance is occurring. Several tools (for example VERITAS i$^{3\text{TM}}$ and Mercury Performance Center) have very specific and intimate knowledge of various applications and architectures and can now make specific recommendations as to how these problems can be resolved.

If this first area for performance monitoring is thought of as being long and shallow across the application, the second area is short and deep on specific items, usually the servers and the network. Identifying that there is an issue with an application component running slowly on a particular server is only half the story. What else is running on that server, and how does that affect the target

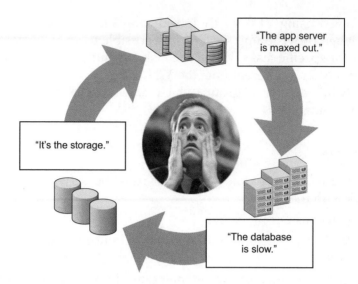

Figure 10.5 Stop the blame game.

application? Tools for monitoring this second category have been around for many years, for example IBM's Tivoli suite of products or CA's Unicenter products. While both pieces of performance monitoring can be used independently, it is the combination that is most powerful and there is currently no combined product that will help correlate data from both the application and the individual resource, so it will still be up to the administrator to interpret the data they have. However, with the two perspectives, the task has become significantly easier than previously, where the finger could always be pointed elsewhere, 'the application is running, it's not my fault', 'the server hasn't changed', 'the network is OK' (Figure 10.5).

It is important that the tools for performance monitoring can be used in production environments. While useful during application development and testing for spotting problems, it is when they are deployed in production that they are most useful for the ITO. SLAs can only be written and agreed to if they are objective, which means that measurements must be taken frequently, in order for the appropriate reporting, and in some cases policy automation, to occur.

Key questions

- Does the organization have multiple groups handling different aspects of application delivery?

- What type of performance monitoring do you require? End-to-end from the user's perspective, or in-depth for individual components, or both?
- Do you need prescriptive analysis or just reporting?
- What else do you need the data to interact with?

10.4 REPORTING

Reporting is perhaps the most crucial component within the utility environment. It is this that enables the consumers of IT to know that they are being delivered what they are expecting in accordance with the SLA, and for the ITO to know that it is delivering it.

Within the utility environment, a great deal of information is collected, much of it pertaining to usage and efficiency. Historical information should be gathered so that trend information can be extracted and interpolated to provide planning from both consumer and ITO perspectives. The ability to manipulate the data in a multitude of different ways is important, linking usage to individuals or groups through the personnel system, or doing something similar for servers through an asset register, should be considered strongly as it will help in creating meaningful business reports (Figure 10.6). However, such linkages need to be kept up to date for the reports to remain meaningful – so this should not be seen as a one-off task. When the environment gets to the point where servers and other resources are being allocated and reallocated on demand, then this too needs to be taken into consideration.

The key to a good reporting tool is that it can be customized to deliver the right content in the right context to the right person at the right time and remain consistent across the organization. Without consistency, it is difficult for the consumer to understand how their requirements have changed over the previous time period. While the consumer reports should be in their own language, for example in transactions carried out rather than i/o per second, those for the ITO should also be in their own language and i/o per second may be far more appropriate, especially when it comes to planning and assessing the alignment of spend to business benefit.

Key questions

- Can the interface be customized to create individual portals for the different users?

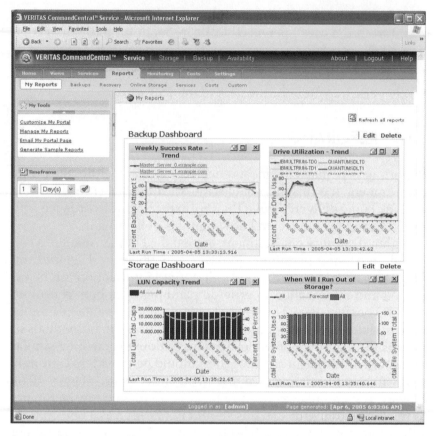

Figure 10.6 An example utility report (from VERITAS Command Central Service™). Reproduced by permission of VERITAS Software Corporation.

- Can reports be customized easily ?
- Can data from additional sources be included in reports?
- Can a linkage be created from personnel or asset management systems?

10.5 AUTOMATION

There are two aspects of automation software:

- workflow for process automation;
- policy automation.

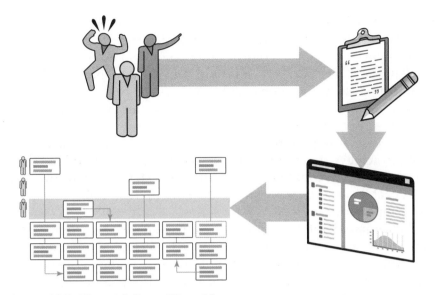

Figure 10.7 The evolution of IT workflow.

Whether moving to a utility computing environment or not, the ability to create standardized workflows for everyday IT processes is a huge benefit. Perhaps one of the most compelling aspects is that it is simple to measure the return on investment (ROI) from such tools. If, before implementing workflows it used to take two weeks to allocate storage, but afterwards it only takes two hours, there is an obvious return. Similarly, if beforehand, only one or two administrators within the ITO could carry out the task, but afterwards any operators can do it, then there is an obvious return to be considered.

Creation of workflow (See Figure 10.7) starts with the ability to write down the steps required. After this has happened, it is important to be able to codify it and take the resulting diagram back to all the individuals concerned to check that the process has been captured correctly. Having seen the overall structure, there is then the possibility to optimize the workflow and bring in best practice. Often, best practice will have to be adapted to the ITO and the way it works, but choosing a product that delivers a number of templates from which to start, speeds the customization. Finally, the workflow needs to have steps automated. It is important that there are *break* points in the process where there is the requirement for human intervention. For example, when allocating storage, while it might be possible to automate completely the allocation, the chances are that the administrator would like to know where the storage is being allocated, whether it is budgeted for, whether it has been approved, and so on. Without this control, automation will not succeed, as it will never be trusted. As trust is built, and perhaps more checks are automated, so further steps can be automated.

The ability to interface with other systems is important, and so the ability to define and add new steps, which can be made part of the process, is important. Simple script-based tools are the easiest way to achieve this, with well-defined processes around how they are used to define inputs and outputs to the steps. As with any tool that enables customization, upgrade paths and the ability to export and import are vital.

Policy-based automation is being able to define an action based on one or more things to be measured meeting a certain set of conditions. Correct use of policy automation can remove drudgery from the ITO and prevent callouts at 2 am in the morning. Often, the policies start as simple rules before growing into complex automations. As with workflow, building trust is important and so semi-automation is crucial, to this end, integration with workflow is good. This way, a set of conditions can be watched for, and when met, a process workflow initiated keeping control within the ITO. For many actions it might be possible to have a single click for acceptance of the proposed action to be taken, for example, if a file system is about to run out of space, then allocation of new space can be automated – provided there is the space available and the project needing it is of the appropriate priority. While the first question on there being space available can be automated, it might be that the project priority needs human intervention. Policy can then be set to increase the space proportionally, for example by 10% or by a fixed amount, e.g. 100 GB, provided the underlying software is capable of the functionality.

Once again, choosing a tool that integrates monitoring with the policy creation and provides template examples that can be customized will speed both adoption of the technology and the ROI.

Key questions

- What type of automation do you require, process workflow or policy-based, or both?
- Are there templates that can be used as examples?

- Can workflows/policies be customized easily for your needs?
- Can new workflow steps be created to enable use of additional tools?
- Are the policy and workflow definitions closely integrated?
- Can you integrate it with your existing ticketing systems?

10.6 CHARGEBACK ACCOUNTING

When utility computing first came to the fore, one of the big-ticket items was *chargeback accounting* – the ability to charge the consumers for what they use. Over the past 12 months, the authors have seen a change from chargeback to a simpler visibility into costs.[4] On the face of it, these appear the same but without the bill! In practice, neither the consumer or the ITO or the business in general is interested in pushing paper money around. That being said, the visibility into the cost of providing a service and, therefore, the cost of using it, are very important in helping align the IT spend with business objectives.

When looking to create chargeback, it is important to find out what type of flexibility is required and to realize the difference between cost and price. The cost is just that, the actual cost to provide and run the ITO, while the price is the indication used to charge or inform the user. The ITO is generally not intended to be a profit center and so the theory is that the price and cost should, therefore, be the same. In practice, setting up an accurate chargeback model to represent this is not practical and so a rule of thumb is often used. This is simple–take the salaries of the people within the ITO, include other running costs, such as electricity usage or rent, and this forms part of the fixed cost of providing the service. Then, take the cost spent on new hardware and support for old hardware and band it together to create the variable costs per service template. These numbers can then be split into a variable usage rate. It might be that inflation is used, along with planned capital expenditure for the year ahead, to give more realistic numbers for the upcoming year to improve the estimation further. Keeping the calculation as

---

[4] The exception here is with outsource firms, who are interested in apportioning costs down to the individual consumers and ensuring the price that they charge does cover all costs.

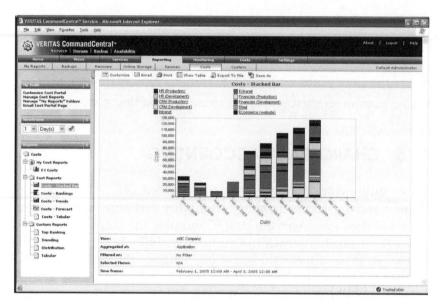

Figure 10.8 Example of a chargeback report (from VERITAS CommandCentral Service™). *Reproduced by permission of VERITAS Software Corporation.*

simple as possible, but at the same time maintaining a realistic value, enables two important aspects of chargeback accounting. First, the price shown will reflect a real number – it is too easy to work in abstract (but proportional) units and have the consumer not realize the 'value' they are getting (see Figure 10.8). Secondly, from a planning perspective, it is important to keep the price constant for a year and cover any fluctuation in hardware prices. For example, the cost of a server will decrease during the year, but the price for the service, whether it is on a server bought at the beginning of the year or one at the end, should remain the same.

Over time, after the value of spending time doing this work is realized, the granularity can be improved and the calculations become more complex. At this point, and provided the resource monitoring is capable of such, it is possible to look at items such as CPU or network utilization. However, it is important that when considering pieces like this, the ITO does not end up being seen footing the bill and paying for underutilized resources. For example, if a consumer requests a service for half a server but ends up only using 2% CPU, then the pricing has to reflect this. Alternatively, if they were to request 500 GB of storage, but only use 50 GB, changing the pricing to reflect actual usage, rather than allocation, would result in the ITO covering the cost of something that it cannot reallocate or reuse.

Key questions

- What purpose are you trying to achieve with chargeback, true chargeback or just visibility into costs?
- What is it that you want to measure? Are you able to collect the right metrics within an appropriate time period?
- Can the chargeback model cope with both fixed and variable pricing?
- Can the chargeback model be created simply and easily by grouping assets into bands?
- Can the model be refined to improve the detail at a later stage, or will you have to start from scratch?

10.7   SERVICE LEVEL MANAGEMENT

Within the utility computing, service based, ITO it is the service management process that is required to bring all the pieces together (Figure 10.9). When looking at software to carry this out, various different roles need to be considered.

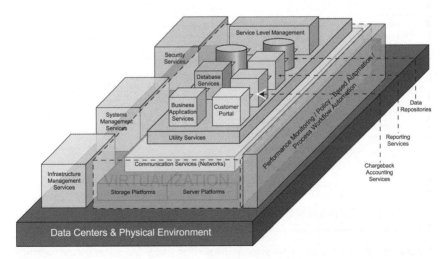

Figure 10.9   Piecing it all together for service level management.

Initially, it will be the administrator who, working with the consumers (and probably vendors), will need to design the services and the levels at which they operate. Extensibility to be able to include new resources into service templates will be important, as the vendors will not necessarily know how resources will be used and which metrics may be used as service level objectives (SLOs). Ultimately, it should be easy to create the templates with associated metrics and pricing information for use in a service catalog. Other information relating to support should also be capable of being entered, so making the service self-explanatory. The goal should be that the consumer does not have to call the ITO for any additional information, and a compromise on the level of detail required to do this will take time to achieve.

From the consumers' perspective a customized portal showing their deployed services and the status is probably the most important piece (Figure 10.10). This would include being able to create reports on whether the service level agreements are being met, with

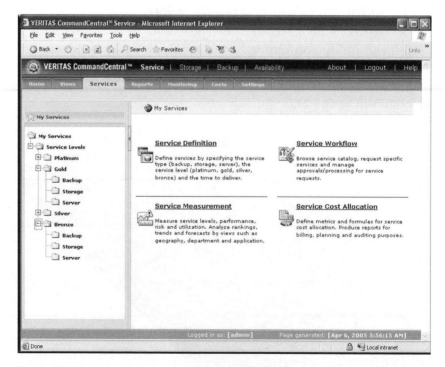

Figure 10.10 Example consumer portal (from VERITAS CommandCentral Service™). Reproduced by permission of VERITAS Software Corporation.

the option of forecasting reports to enable them to look at historical use so that they can plan for the future. Visibility into costs and, therefore, the notional cost associated with their usage should also be available and allow the consumer to roll up costs. When a service level is breached, then notification of the breach must be able to be created through whatever means best suits the consumer, be it email or even an SMS message. The other side of the consumer portal needs to be able to request new services from the ITO's service catalog. Customization by the administrator to enable them only to show services that are applicable to that consumer is important as part of delivering only relevant information. The catalog should show upfront the different costs associated with the services available so the consumer can see at a glance, like a restaurant menu, what the options are. When a request is made, other information will be needed to create the service level agreement, such as contact information and any variable parameters; for example, when requesting 'Gold storage', how much is required and which system is it to be attached to? When the service has been requested, is it to be provisioned automatically through a workflow or will it be carried out semi-automatically with help from the operator or administrator?

The other major player who has an interest in service management is the operator, or the person responsible, from the ITO's perspective, for managing the services. A similar portal to that needed by the consumer, but with greater context, is required to give a more complete picture of the IT environment and its utilization. Service status and breaches will need to be highlighted, with the ability to cross launch to other tools to resolve problems if necessary. Reporting on excess capacity or underutilization of assets will be just as important as current utilization.

The other people who will be interested in the new service-oriented environment will be people like the CEO and CIO. They can be thought of as extra consumers, who will want to see reports based on different criteria, for example, splitting the resource usage down by geography or by department. They will also be interested in the cost visibility to see how well IT resources are aligned with the business. This functionality comes from having a flexible mechanism for grouping services and providing reports based upon the context needed. Once more, data should be presented in the format required, rather than in the format in which the ITO is used to working.

Key questions

- Which pieces of existing IT infrastructure need to be incorporated into service level management? If a phased approach, where is the start point?
- Can the SLM system be modified easily to add new resources and service level objectives?
- Can service templates be defined and integrated with an automatic provisioning, deployment or process workflow system?
- Is the service catalog and service request simple to use? Does it allow for general service level agreement information to be incorporated as well?
- Can services be grouped such that the CIO can see what is needed to get an overall picture, while the database administrator only sees the services they requested?

10.8   HARDWARE

While this chapter is primarily about the software for utility computing it would not be complete without mention of the hardware. Hardware is at the center of the reference model, see Figure 10.11.

For most organizations, being able to make a completely new selection of hardware is not an option. Legacy equipment, both hardware and software, along with long-term purchase agreements with vendors, make it impossible to change just because of the switch to a utility IT model, and so an evolution will be required as part of the transformation program. This also means that the introduction of new hardware platforms can be accommodated.

For initial projects the ability to treat it as a green field project will make it significantly easier to measure the return on investment, as well as setting the right expectations across the rest of the business. This can also allow for the building of relationships with new suppliers with whom the risk can be shared.

While blade servers and interconnect technology, such as Infiniband, can be used in a utility computing strategy, they are not prerequisites. When choosing the services that are to be delivered, it is important not to get carried away with all the new technology that may be applicable. If the project requires tens or hundreds of servers and they need to be provisioned and reprovisioned regularly, then

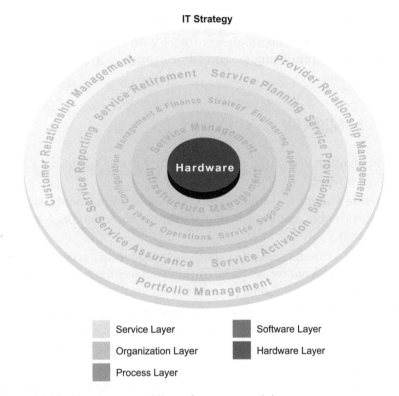

IT Strategy

Service Layer
Organization Layer
Process Layer
Software Layer
Hardware Layer

Figure 10.11 Hardware and the reference model.

a blade-based solution may be the best suited for job. Technology for technology's sake is no longer an option – it is too expensive, not necessarily to buy, but to manage.

Key questions

- Do you need to reuse all the hardware and software that already exists?
- Which pieces do you need to retain? Can you sandbox them?
- Can you go to new hardware (and/or software) vendors for a green field project?
- Could the service be introduced as a green field project, where new hardware, albeit from the existing vendors of choice, can be purchased and used?
- What timeframe are you using for the project? Is it ambitious enough?

- What metrics are you going to use to measure success?
- Do you need to renegotiate contracts with suppliers to support your utility computing project? For example, to share risk, see Section 5.6.

10.9 SUMMARY

- Virtualization is the underpinning of any utility computing technology strategy.
- Software should work in a heterogeneous environment to enable the introduction of new server platforms without having to retrain personnel.
- Centralized management of both heterogeneous server and peripherals, such as storage, is an important facet for future-proofing and allowing for a choice of vendors.
- Selection of tools to support a utility computing initiative needs to be done in order to support the new processes that have been defined in the transformation program – not technology for technology's sake.
- Hardware should not be seen as the starting point for software tool selection.

REFERENCE

Massiglia, P. and Bunn, F. (2003) *Virtual Storage Redefined: Technologies and Applications for Storage Virtualization*. VERITAS Publishing.

Part Three

Implications of Utility Computing

'Hoc erat in votis'[1]

Horace

The final part of this book examines the implications of utility computing. While a methodology for utility computing has been given to help the IT organization transform into a customer-focused, service-oriented delivery mechanism, the implications of the transformation are much wider. Examining the cultural implications for the business and looking at adoption strategies and the relationship change required with suppliers will help reduce the pitfalls. Finally, a look at 'what next'; utility computing is a paradigm shift in the way that IT aligns with the business, but there are other ideas on the not-too-distant horizon which may cause another change in the future.

---

[1] This is what I wanted (or, literally, this was among my prayers).

11

Cultural Implications

11.1 OVERVIEW

As other sections of this book have indicated, a utility computing transformation will change and influence far more than technology design and IT-related process. The very nature of the utility model will often have a profound effect upon several areas of organizational culture. Our experience has shown us that these can include any or all of the following:

- IT service provider (ITO)/business relations;
- IT department focus and code of conduct;
- IT/business alignment;
- business attitudes surrounding issues of data security;
- the perception of an IT department's success in service delivery;
- the nature and type of interaction between business users and the IT department;
- funding models and financial implications associated with IT investment;
- the scope and reach of IT organization's responsibilities;
- the nature of business relationships with partners;
- IT governance.

In this chapter we will explore some of these potential effects concerning culture, as well as methods of mitigating risk in these areas.

Delivering Utility Computing.   Guy Bunker and Darren Thomson

It is important, whilst transforming any type of service in a radical way, that organizational culture is influenced in a positive way. Our experience has shown us that utility computing projects are most successful when effort and time is spent ensuring that this is the case. Also in this chapter we will explore methods of effective corporate communication, a cornerstone of successful cultural change.

As a starting point in this area, our advice (as mentioned in other chapters of this book) is to ensure that your utility computing project has executive-level sponsorship from the start. A positive 'buy in' from the CTO/CFO/CIO (or other) senior executive will ensure that a cultural change and communications strategy can be driven 'from the top'. This is critical since there is always a danger otherwise that (despite your best efforts) project stakeholders do not take the utility computing initiative seriously enough or (even worse) do not commit the necessary time, focus or resources to make the project a success. Everything should be driven by a well-formed and clear business plan. The plan should articulate tangible business benefits and good reasons for embarking on the road to utility computing. Business plans are discussed elsewhere in this book (see Part Two).

11.2 WHAT TO EXPECT, GETTING READY FOR TRANSFORMATION

In this section we will explore some examples of organizational cultural change that the authors have seen occur during utility computing transformations. Within each example, we will pay attention to the following:

- cause and effect of the cultural change;
- risks associated with the change;
- effective methods for introducing positive change.

11.2.1 Alignment of IT and the business

The relationship (or lack of one!) between the business and the IT organization is often one of the main drivers behind a utility computing initiative (Figure 11.1). One of the promises of a 'customer-centric' IT model (like utility computing) is that it should help to

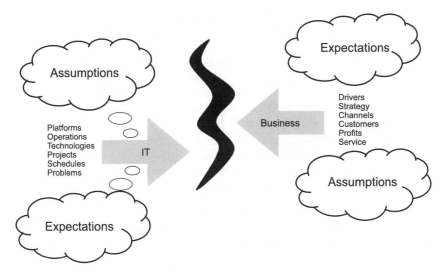

Figure 11.1   The IT/business 'stand-off'.

align business requirements with IT services. Additionally, the theory goes that through this alignment, it will then become easier to measure the value that the IT function brings to a business.

In the years subsequent to the Internet revolution (and subsequent 'bubble burst'), many CFOs and business leaders have become very much more focused on, and aware of, the annual funding associated with IT and technology-based projects. This has led to (in many organizations) an almost obsessive questioning of where the IT bucks are going. It is now fairly widely recognized and appreciated that a closer relationship between the IT and business functions of an organization should bring benefit in terms of alignment (and, following that, cost justification).

By definition, utilities are very *customer-centric*. As has been mentioned previously in this book, utilities cannot survive without an intimate knowledge of, and close relationship with, their consumers. So, it follows that the utility computing model (i.e. recognizing the business user as being a true 'customer' and refocusing the IT department to act as a true service provider) should prove to be an excellent catalyst and enabler to this effort.

As we have already mentioned, a key to this type of relationship change will inevitably be the presence of an executive sponsor. Having the CEO/CIO/CFO fully supportive of your project and familiar with its objectives can prove invaluable in managing potential (and real) conflicts that can arise in a complex transformation of this kind. On many occasions, we have witnessed an executive sponsor dispel conflict by asserting authority and helping the project to reprioritize and refocus, by helping the project team to take a step back and look at an issue from a different perspective, or simply by reinstating the importance of project success and urging the conflictive parties to agree to disagree and move on.

Having experienced some successful (and a few unsuccessful) transformations of this kind, we have noticed that a few fairly basic principles can help smooth the transitional IT/business relationship as the move to utility computing is undertaken. Many of these principles may seem obvious to the reader. However, our experience is that these 'people-oriented' best practices can often be demoted or ignored completely by a technically oriented project team and this, in itself, can prove to be the downfall of even the best technical solutions. This also raises the importance of having exceptional program and project management specialists available for your utility computing project. These types of individual will often have experience of maintaining calm through complex and potentially disruptive transformation.

So, here are some examples of principles that can help greatly as you manage a changing IT/business relationship. The list is by no means exhaustive but captures pretty well the lessons that we have learned in the past couple of years:

- Always secure an executive sponsor and keep them appraised of all major project issues. Allow the sponsor to be vocal about the value of the utility computing project internally, and ensure that they are made aware of potential problems before they occur.

- Create a 'virtual' team of individuals (no more than ten) from both the business and technical parts of the organization. Make the chairing of regular meetings between this team (a Utility Computing Project Board) the responsibility of the Utility Computing Program Manager. Hold meetings between this team as regularly as is necessary to flatten out issues as they arise. Make the meeting compulsory (by order of the executive sponsor) and ensure that a healthy cross-section of organizational roles is represented at the sessions.

- As part of the utility computing project deliverables, create a document (call it what you like) that details the interface points (both technological and human) that exist between the IT utility and the business unit/s that it services. Include within this document a full list of roles and responsibilities and escalation procedures.

- Spend time with the business and agree in detail how you will be able to prove to your consumers that your service is running well and is providing value to them. If necessary, concentrate (initially) on how their service has improved since becoming a utility. This discussion will normally center around the type of management reporting that is required by the project stakeholders.

- Advertise your project successes constantly! Do not be afraid to share your success stories, particularly as you begin to pilot the new utility service. Knowledge of how good the new service is will bring you more new consumers than any other type of advertising.

- Recruit individuals into your project team who have good vertical domain experience in your business field. In other words, try to bring IT experts to the project who have worked with a business like yours before. Compelling business users to try the new utility will be made very much easier if your staff speak their language.

- Take care of existing critical team members. The new world can appear scary and you do not want to lose them. Emphasis on automation removing day-to-day drudgery, rather than their jobs, is important. Close management will ensure that they will stay, good staff are hard to come by – a good administrator is a good administrator, whether you are running IT as a utility or not.

- Remove people from the project if they are not 100% behind it. Having nay sayers on the team will make the project unnecessarily hard to sell outside the ITO, as well as inside.

- Educate and train all those involved. How are the new processes going to work, who will be the contact, what will happen if there is an issue? All of these, along with any new tools that are going to be used, should have some education or training put around them. Remember to give the big picture as well – just why are you investing in this initiative?

11.2.2   A change to the IT funding model

For many organizations, the concept of a chargeback funding model (often associated with utility computing) is the thing that raises concern over how well or easily the utility model for IT will work for the business. As has been mentioned already, it is important to realize that using utility computing as a model for IT transformation does not necessarily dictate that the IT department should charge for its services using a chargeback model. Indeed, many organizations with which we have worked have elected not to change the IT funding model at all, but to use some of the other utility computing design principles to improve the overall quality of IT service delivered. Of course, organizations that wish to provide utility services to external clients have to implement some form of chargeback model in order to sustain their business.

Before exploring the likely cultural effect of changing the IT funding model, let's look at why this type of change often has to be made during a utility computing project.

Most businesses today fund their IT function in one of three ways. These are:

1. Through allocation of operational and capital expenditure budget at the beginning of each financial year.
2. Through allocation of operational and capital expenditure budget at the beginning of a specific new project.
3. By implementing a simple cross-charging financial model, whereby the IT department charges for services or assets by debiting a departmental profit and loss (P&L) account.

In each of these funding models, the business (either at a departmental level or at a corporate level) has been conditioned to treat IT almost as an expensive (but necessary) 'business tax'. Money is either preallocated, with the assumption that the quantity allocation is appropriate, or an arbitrary amount of money is taken from a business, with the assumption, once again, that the amount is appropriate.

All three of these financial models have proved to be increasingly unsatisfactory in many organizations since they do not allow a business to gain a level of confidence in its method or quantity of IT expenditure. These facts often lead a business to consider a utility computing approach to IT due to its association with the financial chargeback model which, in itself, is generally considered to contain the following characteristics:

• IT services are charged for, based on agreed rates (often documented on some form of 'rate card', see Figure 11.2).

	Tier 1	Tier 2	Tier 3
Throughput	High	Medium	Low
Response time	Fast	Medium	Slow
Availability	99.99%	99.95%	99.90%
Support	24 x 7	24 x 7	8 x 5
Cost per GB	$1.60	$1.20	$1.00
Time to capacity	<500 GB – 48 hours >500 GB – 2 months	<100 GB – 48 hours >100 GB – 2 months	<50 GB – 48 hours >50 GB – 2 months

Figure 11.2 Example of a storage chargeback 'rate card'.

- IT services can be modeled so that they can be provided with compliance to service classes (different standards of service). The service's price will be appropriate to the service class to which it belongs.
- All charging should be transparent to the consumers and automated.
- All financial information associated with the chargeback model is documented fully and regular management reports are available through use of comprehensive reporting tools.
- All charging occurs whilst taking agreed service level agreements into consideration. Penalties (sometimes financial) can be incurred by the service provider if these SLAs are not met.

So, what effect should we expect a move to the utility chargeback model to have on organizational culture? Our experience has shown that, of any of the typical cultural (sometimes political) impacts that a utility computing transformation will have, the effect that the model can have on IT financing is perhaps the greatest. Clearly, if utility computing can reduce overall IT cost, then this is seen as a positive to most businesses. However, having a compelling business case in place (complete with return on investment predictions) is only the starting point in convincing a business to fund IT using a chargeback model (Figure 11.3).

There tend to be four main issues that arise (normally quite a long way into a business justification exercise), which present themselves

Figure 11.3 Win over the cynics.

as inhibitors to the adoption of a chargeback model. We have found these to be:

- A 'this is how we have always done it' attitude from the financial side of a business.
- Cynicism towards the new IT services 'rate card' (how do we know that a 'Gold backup service' should cost 50 c per Kb?).
- A reluctance to adopt the new workflows and processes that are necessary to ensure transparency and smooth operation of the chargeback model.
- External consultants who give 'independent' analysis and then come up with a rate that undercuts that which has been calculated by the ITO. If this does happen, look at whether the service offered matches itself to the business, whether the rate allows for reinvestment and covers all the costs, not just the raw values.

These issues are not easy ones to address. Mainly due to the fact that they have typically been raised by individuals or departments who are not influenced naturally by the IT department. Once again, a senior-level executive sponsor can normally help iron out these issues by acting as a mediator between the various project stakeholders. Additionally, we have found the following guidelines to be of use when issues such as these are raised and threaten to slow down a utility computing project:

- Involve the future consumer of the IT utility in the very early stages of pricing definition. Make it clear to all parties that your service pricing will be based on formalized calculations (using measurements of total cost of ownership and so on). If the business gets the feeling that you have simply plucked some numbers out of the air and applied them to your utility pricing, they will not be confident of gaining value for money.

- Introduce a formalized process for reviewing service pricing on a regular basis (perhaps once every financial quarter). Business people know the power of asset (particularly technical asset) depreciation. They will expect prices to fall as the service gains more subscribers and as assets depreciate.

- Remember that, essentially, by adopting the utility model, a business is probably making a trade-off between control and cost reduction. The class of service model will take some control from the business (they will have less choice and will not easily be able to compel IT to change the way that a service is delivered). Spend time creating ongoing reports that relate the cost of the new utility service offering to the old IT model, which, if things have gone well, should show a considerable cost reduction.

- Be very clear with the business from the offset whether your utility is going to be priced so as to make a profit or simply break-even. Once this has been agreed, do not be afraid to share the models and formulas used to create your rate card. Also, make use of internal accountants (who are trusted by the business) during this activity. This will encourage a sense of trust between the service provider and the consumer (Figure 11.4).

11.2.3 Changing the IT department's code of conduct

Some of the toughest cultural challenges that are introduced through utility computing transformation involve the necessary change in behavior and responsibility of the new utility service provider (the IT department). As has been discussed, this type of change to the way in which IT services are delivered involves an IT department becoming much more 'consumer-centric' and focused

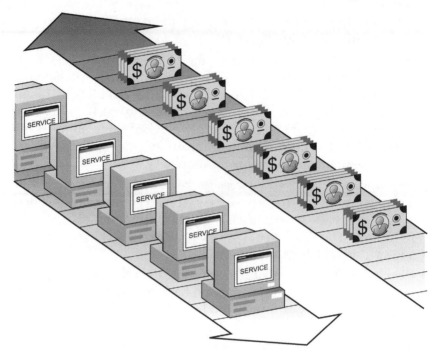

Figure 11.4 Build a strong relationship between the consumer and the provider.

increasingly on creating a positive consumer experience. Broadly speaking, the following attributes are often seen once the IT department is repositioned in this way:

- the IT department treats its business departments as 'consumers', rather than 'funders';
- the relationship between the business and IT becomes more formalized;
- the services that IT provide are based on a standardized 'catalog' of service offerings;
- the IT function and its success in delivering service would largely be based on formal SLAs. These would be measured and reported regularly;
- the IT department would focus heavily on consumer relationship management and new skills and competencies would be developed to support this effort.

These attributes will inevitably have an effect on the way in which an IT department goes about its business. Often, the traditional project-based funding model for IT (an assumption by the IT department that an appropriate amount of funding will always be allocated to the department) can drive a certain amount of complacence within the IT department, particularly concerning the importance of 'the consumer'.

Our experience suggests that an important part of implementing the utility model for IT services involves the coaching and mentoring of existing members of an IT department, in order that they become aware of how their code of conduct will need to change for the utility model to become a success. Ideally, an existing or newly recruited member of the IT staff (with appropriate skills) would be given a 'consumer relationship management' (sometimes known as a business relationship management) role as part of the utility computing project. Understanding the business impact of changes and being able to communicate them to the business is key, and they are also responsible for the training and mentoring of the IT department, as well as the following activities:

- sales and marketing;
- IT/business communications;
- service delivery measurement;
- consumer management;
- consumer entitlement;
- utility service definition (redefinition).

Importantly, we have found that the necessary skills and experience required to fulfill this type of role successfully are, often, not available within the traditional IT department. This is one of the areas of organizational strategy that legitimately requires new talent to be sought out and introduced into the IT organization in order to make utility computing a success. Interestingly, we have worked with organizations that have made their utility transformations successful by recruiting a consumer relationship expert from a non-technical background. It is often healthy to bring somebody into this type of role who can take a step back and focus on consumer-oriented service provision without preconceived ideas about 'how IT should be done'.

11.3   MOVING FROM ASSET OWNERSHIP
TO SERVICE LEVEL AGREEMENTS

In the past, lines of business have owned the hardware and software that their applications run on and they have not been willing to share it. Traditionally, the ITO is given the exact details of what they will be getting and told to run it, back it up and generally manage everything about it. Sizing of the servers required is done based on hopeful estimations of usage two, three or four years out, and when the project has not been successful, the assets have still ended up using up valuable floor space in the data center, consuming power and air conditioning and running at 5% utilization. Not to mention the ongoing support costs for both the hardware and software that is installed. However, because the assets were bought originally by the line of business and given to the ITO, the ITO is powerless to change anything.

Moving to the utility environment with a few selected solutions available appears extreme, especially when the new ITO is supposed to be customer facing. Overcoming the issues of asset ownership and coaxing consumers to rely on service level agreements will be the hardest cultural change that the organization faces. The benefits for both sides are great, but the perceived risk is as well. The ITO needs to be involved in new projects as early as possible, so that they can be there to recommend target solutions based on the template solutions that are provided currently. Accurate costing is important and will help the line of business to make the choice. The other key selling point is mobility. By putting in a tiered approach to the solutions and offering migration from one tier to another (down as well as up), then the cost associated with bringing a new business service online can be reduced by staging it initially on a lower-cost (lower-performance) platform, and then in time, depending on how successful it is, it can be moved easily to something more performant (Figure 11.5). If a service level agreement (SLA) which covers performance has been prepared, then this can be implicit to the agreement. However, at least for the first few SLAs, it is useful for the ITO to be able to show the path that a new project might take.

The ability to move an application from one service level to another depends, not only on process, but also the underlying technology that virtualizes the environment, making the transition transparent. Enabling a line of business to grow its application

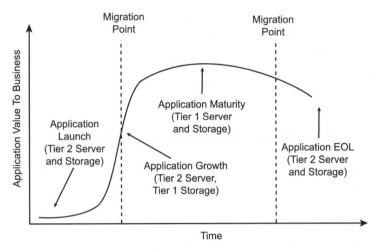

Figure 11.5 Application value over time.

environment in line with demand by changing the level of required services is a significant benefit as it reduces the initial costs significantly. However, it will take one or two projects for it to become accepted as a real capability and, therefore, a value over the traditional asset ownership approach.

11.3.1 Who's in control?

When individual lines of business specified and owned the assets on which their applications ran, it was very apparent who was in control – the lines of business. The ITO was a support and maintenance department whose fault it was whenever anything went wrong. In the new model, with the ITO specifying supported solutions it can appear that the lines of business have lost control over their IT requirements. This would be the case if the ITO decided to dictate what was and was not allowed. However, the success of the ITO is driven by how well it understands the needs of its customers and provides services for them. Providing well-known, and therefore well-managed, templated solutions rapidly in response to business needs actually puts the lines of business in better control of their business – rather than having to concentrate on IT. From the other side, the well-known solutions that can be deployed rapidly and effectively put the ITO in control of the business's IT assets and enable them to use them most effectively. Finally, SLAs ensure that there is a tie with accountability between the ITO and its customers. Better management enables the CIO to have a clearer picture of where IT is being most effective (bangs per buck) across the business and successfully to plan for future investments.

11.4   EFFECTIVE CORPORATE COMMUNICATIONS

We felt that it was important, whilst discussing cultural implications, to say something about effective communications strategies and their importance to radical IT transformations, such as utility computing.

When looking back at all of the problems experienced due to cultural change, our experience has shown us that the root cause of most can be traced to poor internal communication. Once again, the typical IT department does not tend to spend much of its time concerning itself with effective consumer liaison and tends to be more 'inwardly focused'. This can cause problems, particularly as organizational culture starts to transform as a result of a utility computing project. Changes in process and technology cannot be made to work, we have found, without the willing compliance of internal clients (utility consumers) and IT staff. Getting people to change their way of doing things requires a range of communications activities to be carried out in order to generate both initial 'buy-in' and, ultimately, acceptance of the change as a whole. An effective communications strategy (Figure 11.6) can only help a transformational effort, and focus and time spent on this area can pay serious dividends to utility computing projects, particularly during early adoption and piloting of the utility computing model.

In creating and implementing an effective communications strategy, there are a number of key recommendations that we would make. First, the IT leadership team (in partnership with key business sponsors) should agree and articulate a utility computing strategy to all levels of the organization as a prerequisite to the transformational initiative. At the point that the utility computing transformation begins, all stakeholders (both business and IT) should be briefed fully on project objectives, timelines, risks, resource requirements and milestones. Senior executives should be involved actively in this communications program in order to give it credibility and gravitas.

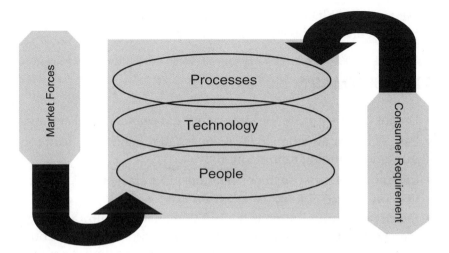

Figure 11.6 The importance of effective communications.

Secondly, the leadership team (both business and IT) should work together to understand the emotions and resistance that are likely to be experienced once the project is underway. The team should think proactively about how each of these situations will be managed and how associated risks will be mitigated.

Lastly, a "best practice" communications approach and strategy should be adopted and applied to the utility computing project, at all stages. This communications approach will need to be tailored based on current knowledge of the business in question, but some general guidelines that can be followed here are described below:

- The IT organization should adopt a consultative approach to transformation and ensure, where possible, that feedback with potential consumers takes place.

- The project team should interact with all project stakeholders with honesty and with full disclosure of project information. Consumers and management staff will find it easier to deal with transformation if they feel they are being dealt with in this way.

- The project team should ensure timely delivery of information concerning the project in order to build and maintain trust. Share both positive and negative project information regularly so that your stakeholders feel informed and involved at all times.

- Make special effort to emphasize good news and to recognize project achievement at all stages of transformation. Build excitement at all levels of the business and compel people to want to be involved and contribute to the project.

11.5 SUMMARY

- Do not underestimate the importance of organizational culture.
- Pay particular attention to IT/business alignment, IT code of conduct and changing funding models.
- Recruit experts in consumer relationship management, invest if necessary to get the right skills on board.
- Effective communication is the key to managing cultural change. Plan your communications strategy in partnership with the business and invest an appropriate amount of project time in communicating with all involved.

12

Developing a Successful Adoption Strategy

12.1 OVERVIEW

In this chapter we will discuss which areas of a typical IT estate are well suited to utility transformation. This analysis will be based on our experience of utility computing projects during the past few years. As we will see, many of today's utility computing projects (certainly the ones that we have been involved in or are aware of) are focused on maturing an IT asset towards the true utility model (as opposed to deploying a completely abstracted IT utility right away). This concept of maturing IT assets is discussed fully within Part Two.

We have found that it is important to adopt a flexible approach to a new IT utility deployment. This chapter attempts to illustrate the choices that can be made in order to plan a project and design a service appropriately. We will also attempt to give an understanding as to why successful utility computing projects have worked well and delivered business benefit to their adoptive organizations.

First though, we will explore how the utility model can be applied to IT using different approaches from the perspective of both scope (how much of the end-to-end IT estate is made 'utility-like'?) and delivery responsibility (which internal and external parties are responsible for the utility service overall?).

Delivering Utility Computing.   Guy Bunker and Darren Thomson
© 2006 VERITAS Software Corporation. All rights reserved.

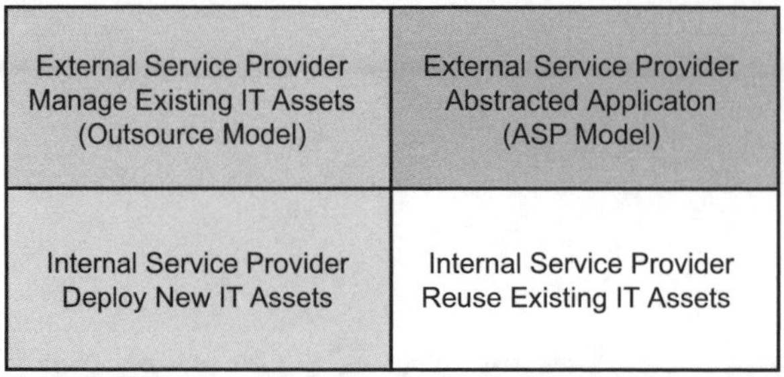

External Service Provider Manage Existing IT Assets (Outsource Model)	External Service Provider Abstracted Applicaton (ASP Model)
Internal Service Provider Deploy New IT Assets	Internal Service Provider Reuse Existing IT Assets

Figure 12.1 Types of utility computing approach.

12.2   TYPES OF ADOPTION STRATEGY

Much of the confusion that surrounds the utility computing concept today is caused by differences in definitions that exist amongst the various technology and IT service providers in the industry. These differences in definition can be attributed mainly to the respective providers' 'means to an end' (what they are trying to gain from their utility computing proposition).

In analyzing and segmenting the various approaches to utility computing that exist in the industry today, we have concluded that essentially there are four types of utility transformational approach. These approaches are summarized in Figure 12.1.

As the diagram shows, an IT utility can be adopted by an organization in a number of different ways. Serious consideration should be given to each of these approaches as a utility computing strategy starts to form. In some cases, services may start in one quadrant and transition to another, for example, a web service might be sourced initially using an external service provider due to internal timing and resource constraints, but then end up being transitioned in-house when resources permit. There are no clear rules for the appropriate selection of approach type, but the following descriptions of these approaches should help the reader to evaluate each of them effectively.

12.2.1   Internal service provider
(reuse existing assets)

The first internal service provider model is put into practice essentially by using the existing (normally remodeled and reskilled) IT department to transform IT assets so that they work as part of an IT utility. This approach does not dictate the procurement of lots of new hardware technology, but will typically require an organization to buy into management software that supports utility computing (workflow tools, virtualization engines, enterprise management consoles and so on). Existing hardware resources are 'pooled' using virtualization technology and, in theory at least, are given a new lease of life (and are worked much harder than before) due to the dynamic resource provisioning model that utility computing requires.

In our experience, this approach certainly represents the most popular choice by organizations that are considering adopting utility computing. Much of the interest in the utility approach exists due to the promise of better hardware asset utilization and improved IT service levels. For this reason, many organizations see an adoption of utility computing by their own IT department (and using existing data center infrastructure) as the most appropriate adoption model. Indeed, many of the organizations that we have spoken to in the past couple of years have seen utility computing as a means of improving overall IT service in order to stop its business from looking elsewhere (i.e. external outsourcers and other IT service providers) for IT services.

Often, areas of the data center made up of very poorly managed and utilized hardware become the targets for this approach (we have seen many organizations target their existing storage and backup infrastructures in this way). It is true to say that these organizations can reap significant rewards (in particular, improved utilization of hardware assets and improved IT/business alignment) having taken this approach to utility computing.

12.2.2 Internal service provider (deploy new IT assets)

The second internal service provider model once again relies on the existing IT department to create and manage a new utility service. However, in this case, new hardware assets are procured since existing hardware is not capable of, or is not appropriate to, the provision of utility services that have been modeled and agreed with business consumers. This situation can often arise if:

1. Existing hardware assets cannot be reprovisioned from their existing function without unreasonable impact to the business.

2. Existing hardware assets are not capable technically of undertaking the functions required by the new utility.

3. An organization has justified the additional expenditure (required for the new hardware) through a return on investment (ROI) calculation of some kind.

In our experience, many utility computing plans that involve this approach go no further than the building of a return on investment (ROI) model, normally due to the fact that 'payback' on an investment cannot be achieved fast enough. However, environments that can use cheap, commoditized hardware components (blade servers, for example) effectively, can often justify this type of approach from a financial perspective.

From the perspective of creativity, it can be quite liberating for a project team to be given the go-ahead to run a utility computing project in this way, since developing a 'green field' environment leaves a lot of room for thinking outside of the box and evaluating leading edge technologies. Caution should be given in these cases, however. We have seen many organizations fail to deliver on a calculated return on investment having become too focused on 'technology for technology's sake'. A proper definition of the business relationships (the service layer of the Utility Computing Reference Model) is critical if this adoption strategy is to be taken.

12.2.3 External service provider (manage existing IT assets)

For some organizations, the outsourcing of IT assets in some way represents the most cost-effective and appropriate adoption strategy for enabling utility computing. Many IT vendors and consultancy organizations now offer services allowing companies to reap the benefits of the utility approach whilst not having to invest heavily in staff or technology.

The first of the two outsourcing approaches essentially uses a traditional outsourcing approach, allowing a service provider to take responsibility for an existing area of IT infrastructure service (system backup or online storage, for example) and to 'wrap' this with their own best practices (in some cases, utility computing) in order to deliver an improved service to the client. In some cases, the IT assets themselves may remain physically within a client's data center (to be managed remotely by the service provider), in others, the physical asset may be migrated to the service provider's data center facility and IT services provided across remote networking links to the client.

Our experience of the success associated with this type of adoption strategy is mixed. Outsourcing in this way will often 'feel' like utility computing since, as a client, all technology complexity will, by definition, be abstracted from you under this type of arrangement. Additionally, the chargeback model often associated with utility computing will have to be implemented by the service provider (in order that they get paid!). However, true utility computing would involve the service provider transforming IT technology, process and organizational attributes in a similar way to that described within Part Two of this book. Organizations that want to reap the true rewards of utility computing should ensure that their selected service provider really has adopted a utility model and has not simply rebranded their traditional outsourcing service in order to breathe new life into an old and commercially flagging outsourcing service.

Another point of note relating to this type of adoption strategy that should be considered is that many organizations that have chosen this method for utility computing find later that they suffer from a loss of control of their own IT services. Tension is often seen to arise between the service provider and the client where, for example, a client's business takes on a new direction which requires a modification to an IT service, but the service provider is not obliged (contractually) to respond to this change.

12.2.4 External service provider (abstracted applications)

The second type of outsourcing adoption strategy involves the external service provider taking responsibility for an entire IT stack, from the back-end infrastructure through to the end user application. This differs from the previously described model, which typically focuses on infrastructure services rather than application services.

Using this model, the client is provided with (typically web-based) applications services (examples that we have encountered have included sales force automation tools and enterprise resource planning applications) across a fast network link of some kind, and is charged by the service provider for the application service based on transaction volume or user count.

The obvious advantage of this type of model is that the client really does not spend time worrying about typical data center issues. A detailed service contract ensures that performance, capacity, scalability and usability issues are handled by the service provider so that the client's business users can simply get the most from their applications. Once again, our experience with this kind of model has seen varying levels of success. Organizations that are considering this approach seriously should start by handing over responsibility for non-critical applications and should explore carefully how issues such as business change will be handled once an application has been outsourced.

12.3 CHOOSING A PARTNER

Making the move to utility computing and delivering various aspects of IT as a service can be daunting, and so choosing a partner to help, at least in the first instance, is important. The transition program from a traditional environment to delivering a service depends upon which service is to be transformed and the size of the business.

12.3.1 Making the first move

Question: *Is this the first service to be delivered?*

The successful delivery of the first service is critical to the ITO, and the partner or partners chosen should be chosen for the specific task. If this is a second or subsequent service, then experience from the previous ones can be used in partner selection. Delivering IT as a service affects the whole organization and there will need to be organizational change, so a number of different skills from the IT project are going to be needed:

- organizational transformation;
- customer relationship management;
- product specialists;
- business analysts.

Many of the skills will be available in-house, but might be difficult to use for political reasons. Business analysts are often available but can create bad-feeling among the lines of business when it comes to putting business value on applications that are used – this has to be done when setting the service level for the various components.

Question: *Where will the primary resistance to change come from?*

Like all good commanders, it is best to have a list of possible scenarios for problems and to attack them – before the service roll out commences. Virtually all the problems will come as a result of the organizational change, and this is where consultants can help to understand the depth of the problems. Interview stakeholders (see Appendix D) and all the key influencers to find out what they think of the idea and what their critical success factors are. Identify possible problem areas and create plans for mitigation.

Many of these types of problems come from people being concerned about change, 'that's the way we've always done it', 'we make a profit, why change now?', 'the ITO hasn't complained before', are all common comments before the first service has been delivered. This is why it is so essential to pick the right first service and to deliver it in a fast and efficient manner. Many people remember business process reengineering projects, which started with good intentions and ended up taking longer than expected and without any obvious benefit to the people they affected. While they might have been successful for the business, the success was never communicated and so people started to fear change.

Question: *What do you want out of this?*

This really is a general question and the answers will depend on who is being asked. It is important to find out what people expected and whether they got it. One answer for the CIO and the ITO should be that the roll out of the service was a success and that they are confident to move forwards with another service.

If this is to happen, then choosing partners who educate and leave behind the skill sets for organizational change is important. Turning the transformational program into a process is critical and then, as with any process, analyzing it and improving the areas which were weak or could be improved should happen before the next project. Trust has to be built up between the lines of business and the ITO, and not just the partner consultants.

Question: *Can one supplier meet all your needs?*

Once again, this is a general question and the answer will depend on what you want from this. The benefits of having a single supplier are obvious, with *one throat to choke* should things go awry. However, utility computing is still relatively new and it is unlikely that you will be able to find a single supplier who can provide hardware and software, as well as the skills needed to manage the transformational program. It is, therefore, better to select the appropriate suppliers and build a team to work together, rather than compromise on one or more of the pieces needed for a successful implementation.

12.3.2   Conflict resolution

When defining services there will be a great deal of pressure put on the ITO and specific individuals to make exceptions to catego-rizing applications and fitting them into particular service levels. The stated goal should be zero tolerance on exceptions. The rea-son for this is simple. Once exceptions are made, or once people think that they can get an exception, then the service will degener-ate rapidly into the existing chaos of every instance being different, and there will be no value brought to the business. The transfor-mational program may be hailed as a success and everyone in the business happy with the change, but then it turns out that this is just business as usual.

12.3.3   New skills for the ITO

Perhaps the biggest change will be in the ITO, which will require new skills in various areas at various times. Over time, these new skills will need to be brought in-house so that the process of delivering IT services can be repeated, but can also evolve as the business does.

The first skill is to understand the business, or a targeted line of business, and be able to relate the way in which IT is used to help; without this, it will be very difficult to determine where service opportunities exist. Once the service opportunities have been iden-tified, then service definition can begin. Being able to call upon prior experience here is useful – what are the best practices for setting up a server utility, how long should it take to provision storage? These questions can be answered by selecting the right partner, and while even best practice will have to be tailored to the business, having a set of templates will speed the process.

In-depth knowledge of the products and tools available will help in defining the services, but it is critical for the people defining the services to be familiar with the ITO and its capabilities, as well as financial agreements that exist, before recommendations can be made.

A critical part of the service is definition of the processes that will be used to run it. These may be manual, or they may use tools to define the workflows. Once more, use of best practice templates is important, but so is understanding how things are currently done so that a mini-transformation program can be put in place. For example, how is storage provisioned today? What does best practice look like? How does best practice fit into the business and how does the transition happen? Education for administrators and operators will be significant, and probably ongoing, while people get up to speed.

12.3.4   Innovation

Utility computing on the face of it is initially about saving money through efficiency and helping to make money through IT agility. One key piece that must not be overlooked is that the ITO needs to innovate to keep the business ahead of the competition. Putting in place a process for innovation and making time for the ITO to investigate new technologies is fundamental to the success of both the ITO and the business. Measuring innovation has always been a tough problem, but it is useful to decide upon some metrics before introducing the process, for example:

- Time to introduce a new service (from conception to first deployment).
- Time to 5, 10, 20 deployments (if the service has been well defined in conjunction with consumers, then it will be adopted more readily. Those services which are not adopted readily should be transitioned out as quickly as possible, or run the risk of becoming expensive to manage as a one-off).
- If replacing an existing service, time to migrate users from the old to the new.

- Number of 'different' services requested per year, versus having something suitable already.
- Number of times a one-off special is required (especially if it is to prevent the consumer going outside the ITO to get the service).

12.4   THE COMPARISON TO OUTSOURCING

The comparison of outsourcing to utility computing should be obvious, they both rely on efficiencies created by the use of best practice and delivery of IT through known good solutions. Outsource businesses could be considered as practitioners of meta utility computing. They have multiple customers for whom they work, and they develop services that they can use for all of them. Even if the environments are different, best practice workflows can be developed and rolled out. Time and resources are available for innovative solution development and investigation of new technology, all to improve the efficiency and cost effectiveness for the outsourcer and not necessarily their customers, who are on fixed terms. Deviation from the original terms often results in large penalties. Business agility and taking advantage of every opportunity that presents itself is, therefore, at odds with the outsource deal, unless terms have been negotiated to allow this to occur.

Of course, tools are the other weapon in the outsourcer's arsenal. These enable the outsourcer to keep a close eye on costs and efficiencies in order to ensure that they stay one step ahead of their customers and suppliers. New tools, such as those outlined in Chapter 10, enable the ITO to become as well equipped as the outsourcer, and even when an outsource strategy is pursued, understanding of the tools available can help in understanding changes in the environment.

12.5   SECURITY

Part of any successful adoption strategy is a close look at the impact of the utility environment on security. The new environment, enabled by virtualization technology, will become in many ways more complex, with applications moving from one machine to another based on performance requirements, data migrating between different levels of service and, therefore, across different pieces of hardware, servers running Linux and acting as web-servers one moment being reprovisioned to be Windows servers with a database, and so on. The more agile the environment, the more closely security needs to be scrutinized.

The first task is to undergo a thorough review of existing security policies and examine where changes need to be made. Assessing the risk and looking for potential threats and vulnerabilities is essential. The subject of security in general, and the various attacks that can be made, are too great for this book, so we shall just look at the areas specific to utility computing.

First, the new agile environment is positive in many ways. The ability to move applications and data on demand means that issues relating to 'acts of God', for example earthquakes and blackouts, are decreased (provided there is a disaster recovery plan in place with a secondary site). By enabling all applications and data to be mobile, rather than just a select few, the impact of catastrophic failure will decrease. Likewise, the impact of individual pieces of equipment failing will also decrease. Risk analysis needs to be carried out regularly to guarantee that critical components are covered, for example multiple paths to storage need to be part of the standard architecture, so that if a switch were to fail it would not affect all the servers and storage using it. When storage was attached to one or relatively few machines, understanding the connectivity and impact of failure was relatively quick and simple. The advent of the SAN made this task significantly more complex and time consuming, such that a tool is required to carry out the topological mapping automatically and make the process of risk analysis less tedious.

Greater interconnectivity and flexibility also increases the impact of a malicious attack. Imagine that someone broke into the system that controlled the utility environment and could provision and reprovision both servers and storage at random, the event would be ruinous. Intrusion detection tools must be deployed. The tools selected for monitoring and managing the environment must be capable of giving a complete audit trail, which can then, in turn, be monitored for strange behavior. Tools should also be selected such that the individual does not have to log in as 'root' or 'Administrator' in order to run them, so that auditing is a worthwhile task.

Automation through workflow and policy helps in reducing security risks, especially those that happen by mistake. Badly designed policies or workflows could increase vulnerabilities, so rigorous sandbox testing needs to be carried out before deployment in the production environment. Being able to use a simple user interface to carry out complex tasks is a boon to operator productivity, but it also means that people can make errors equally quickly. How many times have you dragged an item and put it into the wrong folder? Once more, audit trails are a necessity, so that every change can be monitored and, where possible, actions can be undone.

The new environment enables increased utilization by allowing resources to be shared and reprovisioned. In the case of reprovisioning, security can be breached by a server and/or storage device not being properly wiped clean before the new owner gets access to it. Data files should be removed and the disks reformatted if business-sensitive data has been stored on them. It is good practice to have this as part of the reprovisioning and security policy. Tools for provisioning servers and storage can be used to ensure that this happens.

When a server is shared, especially when the application may have been moved there automatically, there is also the possibility that data access may be compromised. Access control and applications that do not run as root or Administrator can alleviate this problem; however, care should be taken that when storage is moved from one system to another, the access rights are suitably enforced.

12.5.1 Multitenancy

When setting up services, there will no doubt be times when customers explicitly do not want to share servers or even storage (and in the authors' experience, sometimes even backup tape libraries!), i.e. there must be no multitenancy. This seems at odds with the whole utility computing paradigm, which advocates shared infrastructure and standard technology to enable better resource usage and improved efficiencies. However, in this instance, the price charged must reflect the true cost of the service provided and the customer needs to be aware of the change in price due to their specific requirements. Efficiencies will still be made as it will be a standard configuration with standard processes to manage it, and under no circumstances should this be seen as an excuse for a one-off (see Chapter 14).

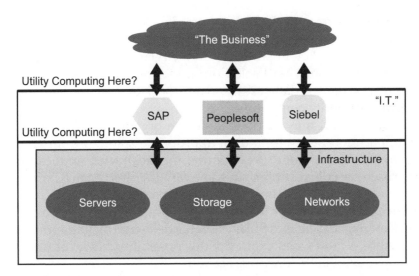

Figure 12.2 Where to implement the utility model?

12.6 GOOD TARGETS FOR UTILITY COMPUTING ADOPTION

Our experience of successful utility computing transformations to date suggests that, in combination with the criteria for good 'low hanging fruit' (see Section 6.3), there are certain other technological attributes that tend to lend themselves well to an implementation of the utility model (Figure 12.2).

First of all, it is important to be realistic in terms of how well today's technology will fit and support an IT utility. Generally speaking, we have found that the creation of *IT Infrastructure utilities*, as opposed to end user application utilities, is sensible. This is due largely to the fact that most enterprise applications today are simply not designed for (and will not allow advantage to be taken from) a true utility computing model.

The issues that surround true application utilities are related largely to the methods of software licensing that are adopted by most of the leading independent software vendors today. Software licensing tends to be fairly inflexible and non-dynamic. A dynamic and agile utility infrastructure often makes demands on an application's licensing model that are either not supported by the vendor, or that penalize the client financially (forcing them to pay license fees for all possible target servers, as opposed to just the ones where the application is running 'live', for example). At the time of this book going to press, a few notable exceptions have started to emerge in the applications market, in particular in the relational database world, where vendors such as Oracle Corporation are investing heavily in the creation of agile databases (and appropriate associated licensing models) for utility computing.

In addition, it may be possible for an organization that desires an application utility to find an appropriate solution from an external service provider (i.e. adopting the 'abstracted applications' approach described earlier). In some cases, a direct match between a business requirement and an application utility offering delivered by an external service provider may be found. In these cases, a client can experience increased business value delivered by this approach. Act with caution when exploring this possibility, however (for the reasons stated in the previous section).

Generally though, our advice is to look at creating business benefit by targeting IT infrastructure services with the utility model to begin a journey to a full implementation of utility computing. In the IT infrastructure service world, once again, we have found there to be certain infrastructure elements that fit the utility model better than others.

An organization that is looking to transform an area of infrastructure service into a utility should take the following into consideration:

- Can all necessary hardware infrastructure targeted support dynamic resource allocation already, or by redesign?

- Do appropriate enterprise monitoring tools exist to give deep, real-time insight into IT infrastructure operations?

- Do appropriate reporting tools exist that will allow business reports to be created that give insight into all necessary and scoped service level objectives?

- Do tools exist to allow the metering of all necessary IT assets for financial allocation (chargeback) purposes?

- Does a service management tool exist that will consolidate all infrastructure management, resource allocation, workflow and financial data for use by the utility service provider?

- How will the software interface that allows the consumer of the new utility to submit service requests be created? Is this tool available from a vendor or will it have to be written as a bespoke application?

As has already been mentioned in an earlier chapter, we have found that infrastructure services providing storage resources have proven themselves to be valid targets for utility transformation. In particular, the area for backup and recovery is an excellent target, due to the fact that backup infrastructures (in most organizations) now rely on access to a shared storage infrastructure (storage area networks, shared tape silos and so on), and software tools already exist in the market today to handle automated resource provisioning, partitioning of 'virtual' hardware resources and business level (service level objective) reporting.

Organizations that wish to trial utility infrastructure services internally will often find that system backup and recovery represent an excellent 'proving ground' for the approach and, typically, return on investment (ROI) characteristics can prove to be compelling enough that the business gains an appetite for further utility computing transformation elsewhere.

So, storage is a good place to start (Figure 12.3). This should not, however, preclude an organization from exploring other areas of infrastructure service that may be provided with a utility offering. Much of the time, areas of service, such as server provisioning, network provisioning or bespoke applications, can be 'utilitized'

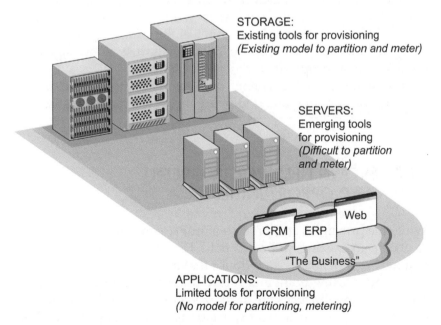

Figure 12.3 Starting with storage.

successfully as long as a realistic scope is defined for the service that is to be offered.

Unsuccessful utility transformations are often caused by the project team simply trying to do too much in the initial transformational project. This is very often the case in 'server utility' projects, where an organization might attempt radically to transform an entire estate of critical production servers that run many differing types of applications and have many interdependencies with other internal and external systems. In the case of the server utility, we have found it to be much more realistic to target, for example, an application development server environment, or a test and quality assurance environment. These types of environment carry far less business risk, but can still often benefit greatly from a utility transformation.

A note to service developers

When looking at rolling out a service, considerations must be made as to how the tools to support the service are delivered. Web based applications, or Java-based applications (which can be downloaded easily and regularly) offer great flexibility, with the key function that they can be updated easily. While pop-ups and other adverts are a nuisance when browsing commercial websites, having the ability to advertise new services, successes and even things like planned downtime is important. This does not have to be as overt as a pop-up, but could be just a 'new' sign and a link to an appropriate intranet website.

If looking to use 'fat' clients, then ensure that you have mechanisms for updating and deploying them on a regular basis.

12.7   BROWN FIELD VERSUS GREEN FIELD OPPORTUNITIES

While the principles and process of utility computing are designed to evolve an existing, or brown field, IT organization and its assets to a new regime, there is no doubt that a green field project can be a valuable starting point and future reference. Not having to work within existing practices means the development and imposition of new ones is easier, customer expectations can be set upfront without having multiple caveats for special cases when referring to legacy systems.

More importantly, a green field project will enable an ROI study to be easily carried out. When evolving an existing environment, it is easy to distort the ROI figures because of all the special requests to deal with existing non-standardized environments. Simple measurements, such as the time taken from request to provisioned storage, can be made in both the brown field and green field projects, and the impact of tools such as process automation workflows assessed.

Once complete, the green field can be held up as the example for other projects and initiatives and used as a business driver-to transform existing environments.

12.8   USING IT CONSOLIDATION AS A STARTING POINT

We did not think it right or proper to write a book about utility computing without saying something about its potential effect on IT consolidation. During our work with clients over the past few years, it became increasingly apparent that some of the theory and method involved in creating IT utilities lend themselves very naturally to the typical IT consolidation project. In the remainder of this chapter we will explore this in some detail and explain how, often, an IT consolidation initiative can be a perfect place to start introducing utility computing concepts to a business.

First of all, though, we will explore the types of IT consolidation project that tend to exist, since utility computing can bring massive value to some but less to others. Exploring how utility computing theory relates to consolidation should help the reader to decide whether their own consolidation strategy could benefit from the utility approach or not.

12.8.1   Types of IT consolidation

The term 'consolidation' is heard a great deal nowadays, and 'the consolidation project' is often cited (particularly within large organizations) as a highest priority that exists within an IT strategy. This is often due to the fact that the concept of consolidation is widely believed to be a major source of cost saving for a business whose IT estate has been growing exponentially in recent years.

Very often, IT organizations (that are driven hard by the business to save money) will embark on a consolidation initiative without first thinking about the types of consolidation that would bring the most benefit to them. The typical drivers that tend to compel IT organizations to investigate the benefits tend to be:

- IT asset underutilization;
- uncontrolled IT asset growth;
- a lack of IT standards (or no standards at all);
- complex architectures (leading to high IT management costs);
- duplication in application functionality;
- poor data-sharing capabilities;
- a desire to reduce the number or size of data centers.

This list is not exhaustive, but is representative of our experiences from consolidation project work during the past ten years.

Interestingly, you will notice that many of the issues listed here are also common drivers for utility computing. There are good reasons for this, which will be explored later in this chapter. What is also interesting is that, in order to approach IT consolidation properly (and with good effect), the IT organization should analyze carefully which of these issues they are really trying to solve. Once this has been decided and agreed with business stakeholders, a *method* of consolidation should then be selected.

We have experienced IT consolidation strategies that attempted to make use of consolidation methods that were completely inappropriate when the true business drivers (that drove the project in the first place) were taken into consideration (Figure 12.4).

So, what different consolidation methods exist and how are each of these relevant to the common business drivers that lie behind a consolidation initiative?

This chapter does not pretend to represent a source of best practice for consolidation but, in the authors' experience, there are essentially three types of consolidation approach, these are:

1. Application consolidation (reducing the amount of applications that exist).

Figure 12.4 Never underestimate the consequences of getting it wrong.

2. Logical application infrastructure consolidation (reducing the amount of hardware asset by allowing multiple applications to share hardware resources).

3. Physical infrastructure consolidation (reducing the amount of hardware asset by resizing it so that it is appropriate to real application requirements).

The term *data center consolidation* is often heard in order to describe a project and its objectives. To consolidate data centers (the amount of actual IT hosting sites or locations) tends to require the application of all three consolidation methods listed above.

A very common mistake within organizations that jump into consolidation without first exploring the true and most pertinent business drivers, tends to be rushed and badly planned attempts at 'logical application consolidation', with a belief that this method will, by definition, lead to a reduction in hardware asset (physical infrastructure consolidation). Often, this attitude and focus is fuelled by excitement (driven by vendor marketing) surrounding software virtualization tools. We have seen many 'server virtualization' projects follow this path through a belief that the logical partitioning of large servers, and the subsequent logical consolidation of applications, will reduce cost significantly. This is often proven not to be the case, mainly because:

- The complexity of managing a virtualized server estate can often introduce cost and business risk that did not exist before.
- Many applications simply do not operate well within logically separate (but physically shared) hardware assets.
- The overall total cost of ownership reduction experienced by having logically consolidated applications is often below the expectations of the business (largely due to point 1).

In our experience, a good and truly beneficial consolidation project is always planned based on some clear and concise business objectives. The method of consolidation (or combination of methods) should then be selected carefully based on these. Here are some examples of clear consolidation project objectives that we have seen:

- consolidate the amount of storage suppliers to two;
- consolidate all disparate databases to run on a single version;
- raise server utilization to an average of 50%.

In each of these examples, the project objectives are simple and, based on these, a project team incorporating engineers, architects, project managers and strategists can work to create a logical and appropriate consolidation strategy.

12.8.2   How can utility computing help?

As we worked with more and more businesses to investigate how utility computing may be of benefit to them, we started to notice that, very often, the business drivers responsible for the instigation of a consolidation project would match the areas that we thought a utility model might improve. In particular, organizations that had targeted large reductions in overall IT expenditure, or that were very concerned about low IT asset utilization, became very interested in the utility computing method. There is good reason for this, and it is due largely to one of the fundamental principles of the utility model, namely, 'classes of service'.

A class of service model essentially allows us to create fairly rigid and well-defined IT services (the parameters of which are captured and abstracted to give a service level, e.g. Gold, Silver and Bronze

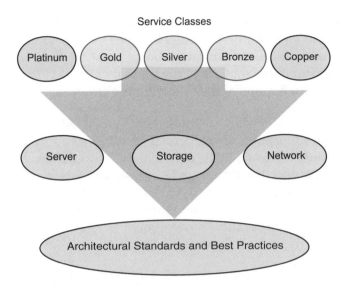

Figure 12.5 Service class definition.

(see Figure 12.5) and then incorporated as part of an SLA). It also promotes a dialog, early in a project lifecycle, between the IT organization and the potential consumer (the business) concerning precise service deliverables such as, for example, performance, availability and scalability requirements. When designing an IT utility (or indeed any IT service using a class of service method) we start by understanding, in some detail, what a service will provide. This is not typically how IT infrastructure has been designed in the past. Our experience has been that this method of IT service design will promote the following characteristics, related to IT architectures:

- appropriate, 'right-sized' IT infrastructure;
- rigid and well-controlled IT standards;
- hierarchical IT infrastructures (allowing applications to make use of appropriate types of infrastructure, based on their performance and availability requirements);
- agility in deployment (speeds time to market);
- scalability.

The last point here, in particular, can drive massive success (in terms of cost savings) into a consolidation effort. Indeed, many of the utility computing projects with which we have been involved have been considered successful purely due to a class of service approach, enabling an organization to design a 'tiered' IT architecture so that the various applications within the business can make use of the IT assets and standards that are pertinent to them.

Consider for a moment the following use case:

An organization wishes to implement a new server estate for use by 200 application developers. Its traditional approach to this project would be to talk to the potential user population about their requirements and then design an IT infrastructure (servers, storage, networking and so on) for a 'worst case scenario'. In other words, all components of the new infrastructure would be designed to ensure that the application development environment would not suffer from any of the following issues:

• poor performance due to increased application load;
• lack of storage space caused by data growth;
• poor network performance caused by increased network based data traffic.

Under these circumstances and using a 'worst case scenario' philosophy, the project team would be forced, in all likelihood, to 'over-engineer' the infrastructure for use by the application developers – 'better to have too much than too little'. Of course, this design philosophy is the root of many of the issues associated with the data center today. This approach to IT design is the cause, in particular, of the underutilization of IT assets (because the assets were designed to be far larger than they needed to be).

Now lets consider how the class of service and utility computing design models could be applied to this project and what the potential effect of this might be. Let's assume that the worst case scenario design approach would see the organization in question having to spend $10 million on IT infrastructure.

Now imagine that, before designing the new application development environment, the project team consulted with the application development staff about the likely IT infrastructure requirements and, in particular, a class of service model. This consultative dialog could, for example, table the idea of a hierarchical IT infrastructure categorized as service classes labeled 'Gold', 'Silver' and 'Bronze' (as shown Figure 12.6).

Service Class	Estimated Cost (Per Server)	Service Level Agreements
Gold (Fast Performance, High Availability)	$50 000	90 Transactions per sec. 99.99% availability
Silver (Medium Performance, Highly Resilient)	$30 000	70 Transactions per sec. 99.5% availability
Bronze (Slow Performance, High risk of "Downtime")	$10 000	30 Transactions per sec. 99.0% availability

Figure 12.6 A simple service class rate card.

If it becomes apparent to the project team, having consulted with the new IT utility consumer (the application developers), that only 100 of the development staff actually require the SLAs associated with the 'Gold' IT infrastructure service class and that 50 of the user population could meet their objectives by using the 'Bronze' standard then this would have a very significant effect on the costs associated with the new project. In this case, a $10 million project could now be delivered for around $7 million and this cost saving could be enabled entirely through use of a 'class of service model'.

This is a fairly trivial example of how the class of service approach can help to consolidate IT infrastructure. Nonetheless, the example is representative of several successful projects that we have seen benefit from the utility computing design model.

12.9   SUMMARY

- Be careful to choose the right overall adoption strategy (and partners) for your initial utility computing project.
- Bear in mind that technology will often dictate what is possible, and keep the scope of initial transformation simple.
- Storage (particularly backup) is often a good place to start.
- Look for utility computing opportunities wherever the term 'consolidation' appears.
- Adopting a 'class of service' model does not necessarily mean an adoption of utility computing.

- The class of service method allows hierarchical architectures to be created and can lead to serious cost reduction (particularly capital asset expenditure).
- If poorly utilized IT assets are a key driver for a consolidation project, then the utility computing method should be considered seriously.

13

Future Trends

13.1 OVERVIEW

Looking to the future is always fraught with danger, ten years ago who would have foreseen the rise of the Internet, email and the subsequent menace of spam? While the Sony Walkman became ubiquitous for music on the move for cassettes and CDs, the mini-disc, with its superior capacity, failed to capture the imagination, but Apple's iPod and other MP3 players caught the imagination of the public and recreated the market – 60 GB of music in something the size of a music cassette, and it will just increase.

The same is true for enterprise IT; within the industry there is always lots of buzz around new technologies and methodologies. Some of these might become niche players, while others will disappear into the ether, very few will make it to the mainstream. Without a doubt, those that will make it big will be driven by customers, rather than vendors – vendors can guide, but customers have the money! This chapter looks to some of the future trends within IT in an attempt to discover what might happen after utility computing becomes mainstream.

The data center of the near future will be based on utility computing processes and practices, which enable the ITO to deploy and redeploy resources as and when needed. The various different tiers will become self-managing based on workload demands (Figure 13.1).

Delivering Utility Computing.   Guy Bunker and Darren Thomson
© 2006 VERITAS Software Corporation. All rights reserved.

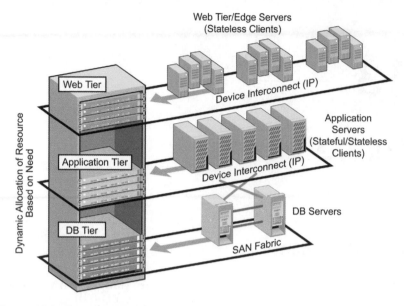

Figure 13.1 The data center of tomorrow.

13.2 STANDARDS

While there are many fledgling standards within the industry, it is up to you, the consumer, to become active in standards generation for them to become successful. Stating which standards would be useful to you and pushing vendors to work together to define and implement them is important. Standards move glacially slowly; even if the definition is quick, it takes years for adoption and even longer before it becomes pervasive in the environment, but they are the key to success in the future. Without them, proprietary solutions will continue to emerge and create vendor lock-in.

13.3 PACKAGED SOLUTIONS

Utility computing as we have defined it is not a packaged solution; it is not possible to walk into a vendor and request 'one utility computing solution please'. However, as more experience is gained by customers and vendors alike, the ability to capture best practice and package it will increase. As standards in various areas surface, process automation workflows will become more intelligent and standardized, requiring only shallow customization for use within the ITO.

The initial packaged solutions will be around backup and online storage, as these are the most mature in terms of technology and deployment within the business. Asset ownership has not existed within the enterprise backup environment for a long time and shared storage is now becoming the norm, tools to enable consolidation of data and usage reporting are also well defined. While an 'instant' solution may never be available, the amount of customization will be minimized.

Packaged utility server solutions for brown field sites are unlikely, but ones for green field projects, especially those based around blade server technology, will occur. A standard software stack can be deployed across a blade farm, virtualizing the storage, as well as the servers, and enabling application mobility. Best practice for application resource tuning will start to emerge in the next two to three years, which, in turn, will lead to more packaged solutions.

The possibility of a service such as disaster recovery being offered as a utility-style packaged solution is unlikely due to the differences between businesses. However, a set of best practices and metrics for SLAs will appear over time.

13.4 SERVICE-ORIENTED ARCHITECTURE

As this book goes to press in 2005, the new buzzword is SOA, or service-oriented architecture. The idea is simple, instead of developing an entire application, abstract functionality into services and then build applications using both internal, and more importantly, external services and coordinate the data between them. A set of standards has been developed to enable the interoperability between applications and the functionality they provide.

Web services are often mentioned in the same breath as SOAs, so much so that it often appears they are the same thing. They are not. SOAs are a principle and web services are just an implementation. Other technologies exist, around which an SOA can be built, such as CORBA, J2EE and .NET. It is the adoption of a limited number of standards that will make the difference, without them, the ability for one service to interoperate with another will not happen.

SOAs are designed to enable the creation of a flexible environment, where changes in business needs can be reflected quickly in the applications that serve them. In the past, integration of new functionality or data feeds would take a long time as there were no standards, but SOAs will, in theory, simplify this. Unlike an object-oriented approach, where the data and the methods are bound together, services receive a message with the data in it and process it accordingly. This leaves the implementation of the service open and so legacy systems can be brought readily into the new architecture by wrapping a service around the functionality that is required. Reuse comes from being able to reuse the functionality (service) offered, rather than the object, which implies both data and functionality.

Implementation of SOAs is, therefore, more about design principles rather than a concrete implementation. From that perspective, the same is true for utility computing, the devil is in the details. Understanding exactly what a service will do is important – how does it deal with units and time zones, and, perhaps most importantly, is there a service level agreement as to the availability of the service? If your business depends on a third party service in order to run a business critical application, you need to be assured of its availability when you want to do business.

All that having been said, there is certainly a good business case for delivering functionality as a set of services that can be plumbed together easily using technology such as XML and, even if it does not turn the world into one large marketplace of services, within an organization, the benefits of improved interoperability and faster integration will be felt.

13.5 VIRTUALIZATION

Virtualization is probably the most overused buzzword of the last five years; initially associated with storage, it is now associated with just about everything from the network to the server and applications. The benefits of virtualization are obvious, being able to isolate applications or data from the physical hardware enables greater flexibility in the environment. Tools have grown up, not only to help the ITO manage virtualized environments, but also to hide new virtualization technologies. For example, storage may be virtualized on the host, in the storage itself or in the storage network, which, in turn, can be in-band or out-of-band. The key to its success has been that the tools have grown up with the technology and when a new version comes along it is supported in the tool, so that the ITO does not have to learn something new in order to make the most of it.

OS virtualization is being done by all the major OS vendors, with a number of smaller software companies competing in the Linux space. Furthermore, Intel recently announced that it is going to put virtualization onto its processing chips. In doing so, it will undoubtedly create a set of de facto standards for managing virtual machines, which other vendors will adopt in the future.

13.6   THE END OF APPLICATIONS AS WE KNOW THEM?

The application service provider (ASP), or rather the idea behind the ASP, rose to prominence in the late nineties, only to be knocked back down in the .com crash of 2000. The idea was simple, why should you spend time and effort on creating an infrastructure and environment to run business applications if you could pay as you go and have someone else handle the IT infrastructure headaches? The idea was a good one, but ahead of its time given the economic environment. Several questions were unanswered around high availability and security – did you want to bet your business on IT that may or may not be there? Did you want to put your data on hardware that could be shared with your competitors?

Now, many of those questions have been resolved and we are starting to see the rise of the ASP once more. Of course, not everyone is going to rush to the ASP model (we believe it is similar to outsourcing), but that customers are going to expect vendors to support a variety of alternative licensing models, so that they get pay-as-you-go-type pricing.

Traditionally, software has been sold at different prices depending on the *tier* of the server on which it will run, so you would pay more for a server with eight CPUs than one with just two. CPU pricing comes in a variety of flavors, and the first is to license based on the number of CPUs, so if you have eight, then you pay four times as much as if you had two. Importantly, this means that you could, in theory, take the license from the eight-CPU machine and instead have four dual-processor boxes running the software. In the utility computing environment, the ability to deploy and redeploy licenses in a flexible manner will be critical if the business is to improve the link between price and value to the business. The other option in CPU licensing is paying for actual CPU cycles used. Sun Microsystems has made a big play of charging $1 per CPU hour used. This sounds simple, but in reality, when there are many tiers to an application, setting up and monitoring CPU usage across the entire system is time consuming for the normal ITO. Longer-term-based licensing is another alternative, where applications may be used for a number of months, or even years, rather than *forever*. More likely, we will see licensing based on something more meaningful, such as sales transactions carried out, rather than CPU hours used, or even on a more business-oriented metric, such as percentage of seats filled.

Other licensing programs will be introduced, for example storage, be it online or offline, based on the quantity of storage under management. This, again, will simplify licensing to abstract out the hardware resources used. If you want a petabyte on a single server, or a terabyte on each of 1000 servers, the pricing can be the same. Once again, this will give the flexibility to move storage around without it impacting the associated license cost.

Note

The change in licensing to a system based on usage will not necessarily make it cheaper for consumers. Like hiring equipment, if the term is short, then it is probably beneficial to hire rather than buy. There are other advantages to the model, not least, not having to worry about infrastructure as that will be 'someone else's problem', and similarly, the flexibility to start or stop, increase or lower usage without too great a penalty.

13.7   GRID COMPUTING

In Chapter 2, we compared utility computing to grid computing and determined that from 30 000 feet they were different (due to the applications that are currently running) but, from three feet, there were a lot of similarities in the required infrastructure. Fully realized, the grid will link *all* available computing resources together and manage tasks across the entirety (Figure 13.2). In a sense, grid computing represents the ultimate utility – users would connect to the grid, use whatever resources they required and

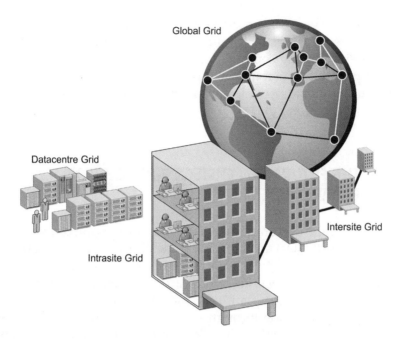

Figure 13.2 The grid.

disconnect. To this end, the grid would be similar to today's conventional utilities – turn on the tap and get as much water as you need; make a telephone call and talk as long as you like and only pay for what you use.

This level of utility computing remains in the future, although there are signs of progress. Collaborative working and resource pooling are in operation today in the pharmaceutical, financial, automotive design and animation industries. But even there, applications that can take full advantage of grid computing are rare; there will need to be a seed change in both education and development tools to bring applications that exploit the grid to every enterprise. This will occur over time and intra-company grids will grow. The likes of web services as an interface will enable inter-company working and will hide whether there is a compute grid behind the interface. In many ways, this is the way it should be; utility computing is about not concerning the user with the implementation, and the grid is the same – as long as the service is delivered as expected it should not matter if it is delivered on a single supercomputer or a thousand networked PCs.

Thomas Watson, IBM's chairman, made a famous observation that there would only be a world market for five computers, the time 1943. Maybe Watson was not so wrong. If 'the computer' of the future is actually a distributed grid of computers managed as a single entity, making compute services available to anyone for running applications, then maybe, just maybe, 20 years' time will see a handful of global providers, as is the case with electrical power, telecommunications, and a variety of other utility services today.

13.8 THE FUTURE: AN OBJECT APPLICATION ENVIRONMENT?

Within the realm of storage there is a new concept called *object-based storage devices* (OSDs). In essence, the idea is to create self-managed storage for storage networks in a heterogeneous environment. By moving low-level functions, which are traditionally implemented differently on different platforms, into the storage device and accessing the information through a new object interface (rather than SCSI or IDE) functionality, this idea can be extrapolated to turning whole application environments into single managed objects.

Utility computing, with its principles of standard template solutions and automation, could wrapper a multitiered application environment and enable it to be deployed and managed as a single object. If IT is to become a true utility, then it needs to become transparent – we do not worry about how electricity is generated, we plug in our device and expect it to work. Similarly, we should not have to worry about how the IT infrastructure is laid out, we should be able to just plug in the application and it will work.

13.9 SUMMARY

- Ultimately, standards will be needed for full realization of utility computing, from protocols allowing applications to intercommunicate, to management of the utility infrastructure. These will take time to mature, both technically and politically, but user adoption of the utility paradigm will hasten that maturation.
- Virtualization will continue to be important in IT environments, as well as the underpinning for a utility computing strategy.

- Licensing models will change, and applications will become pay-as-you-go or usage-based. This will not necessarily make it cheaper for the customer – just more flexible.
- Utility computing as the concept is understood currently may be just a stepping stone towards an even more flexible information processing environment, commonly known today as *the grid*.

14

Afterword: Avoiding the Pitfalls

14.1 OVERVIEW

This book has presented a methodology for utility computing that the authors have used successfully in transforming the IT organization (ITO) from a traditional environment to the more dynamic service-oriented one that is needed for the enterprise of tomorrow. Far from being something that only big corporate enterprises can consider, much of what is contained here is applicable to the small to medium-sized enterprise as well. While the budget for tools may not be as great, the principles and methodology are equally applicable.

This chapter looks at some of the pitfalls that can occur and how to avoid them.

14.2 RETURNING TO CHAOS

Even when a complete project plan to transform the ITO into a state-of-the-art, service-driven organization has been completed and the first few services have been rolled out successfully, there are still areas where caution needs to be exercised.

Delivering Utility Computing. Guy Bunker and Darren Thomson
© 2006 VERITAS Software Corporation. All rights reserved.

14.2.1 The one-off special

All the way through this book, the emphasis has been on creating an IT environment where there are no surprises. Template solutions built using workflows based on best practice built around real customer needs; it makes it sound very simple to deliver on the initial promise of increased efficiency while decreasing costs. The reality will be different, with many lines of business wanting special dispensation for various projects because they are *critical* and so should be treated differently. Here is the first problem that needs to be overcome – dealing with the one-off special.

Unfortunately, precedent in creating a one-off special will open the flood gates to a chaotic environment. It may seem obvious that where cost savings can and will be made from standardization, creating a specialized solution will be more expensive, not just to deploy, but also to manage. This does not mean that it should never happen, it does mean that it should be costed appropriately and charged for. If the line of business understands that deviating from the standard template solutions results in a service that costs two or three times as much, they may reconsider. It is important that the ITO does not dismiss the request out of hand, but is able to back up the statement with some firm figures. The first one-off should be analyzed very carefully so that it can be used as a template for future requests. By putting in place the instrumentation and reporting for an agile utility computing infrastructure, apportioning cost for the one-off should be simplified. Simple questions might be:

- How many administrators and operators do I have that can manage this new environment?
- How long will it take to patch, or upgrade, the OS and applications in the new environment?
- Can I use existing storage and backup solutions?
- Will this environment need to be replicated elsewhere?

If the answer to the final question is 'yes', then the ITO should look at the one-off as potentially becoming a new service template. It is important for the ITO to be able to recognize the difference between a one-off and a potential new service. This will only happen if the ITO remains in contact with its customers and understands both the current and future business needs.

14.2.2   Continuing as before

Rolling out a service will be just that a new service, it will not be a simple switch from the old way of doing things to the new, and herein lies a potential problem. The ITO must remain resolute that after a service has started to deploy, all new requests should use the service. For example, if backup has been turned into a service and a new server provisioned, then the backup of that server should be done using the backup service rather than having it done as it was before. A line of business might not have any of its existing backup being carried out as a service, but this does not mean it should not start as soon as possible. Migrating to the new services should then be made a priority where possible; for things like backup, this is practical to do, whereas moving to a new server probably is not.

As before, allowing a one-off, especially if the excuse is that it is easier, will set a bad precedent and should be avioded. Having IT functions deployed as services, as default behavior and enabling the business visibility into the costs with efficient usage reporting, should speed the adoption of the service rather than hindering it.

If it is more complicated to deploy the new service than it would have been to continue with the old method, then analysis as to *why* must be carried out. It may be that administrators and operators are still unfamiliar with the new way of working in which case additional training should be considered. If, on the other hand it really is worse then the service must change. It might be that processes need to be improved or the entire service needs to be overhauled to become efficient, either way spending time to ensure the service can be deployed efficiently and managed is essential.

14.2.3   Losing the competitive edge

Perhaps the biggest issue that the ITO faces is that all the emphasis is on process and standardization, and while time-to-market is critical in today's competitive business world, it is not the only aspect. IT, in itself, is not a differentiator, how you exploit it is. Staying ahead of the competition is about keeping abreast of technological advances and figuring out how to use them to best advantage by creating innovative new services.

14.3 INNOVATION

Throughout this book, the emphasis has been on creating an IT environment that runs efficiently on standard template solutions and ensuring that everyone complies. But cost is not the only reason to do this. Saving money is one thing, putting the money saved to good use is totally different – and is of primary importance when it comes to the ITO helping the business make money (see Figure 14.1). (Un)fortunately, the world and technology march onwards, and it is up to the ITO to separate the wood from the trees in discovering and exploiting the new technology and using it for competitive advantage.

Creating innovative new services is not just about the technology, all the other factors surrounding a successful service have to be employed. Understanding the new technology and then being able to match it to the business is a start, it is then up to the ITO to distill the business benefits into something their customers can understand and go and sell the idea for the new service. Will the new service be more cost effective? Will it provide improved support and maintenance? Exactly why is this service going to be good for the business? New technology from the perspective of the ITO is *fun*, new technology from the perspective of the business is anything but. There is seldom any time or money available to move applications onto new platforms, so changing platform in the middle of an application's lifecycle will only happen if the benefits are obvious and, where possible, without risk.

Figure 14.1 Innovation to beat the competition.

The other side of innovation is entirely internal to the ITO. Are there new technologies that can be introduced transparently, as a means of reducing cost? In the new utility computing environment, where the ITO provides a service which is driven by service level agreements rather than its customers' asset ownership, it is possible for the ITO to replace or upgrade services without having to ask permission first. Of course, there is the requirement that the SLA must still be met, but other than that, there is no reason for the ITO not to do anything that makes its life easier. Perhaps this could be the introduction of a new storage array that uses SATA[1] disks as a new storage tier; as this has improved performance, it could be used to cut costs simply by migrating storage from other arrays.

The new environment should enable the ITO to foster an innovative environment, allowing senior administrators to look for new technology and then find its applicability to the business. However, there is a danger that every piece of technology is useful to at least one person, resulting in services that are used only once. Worse than this, once an assumption is made that a new service will be useful to more than one person, time and effort can be spent in enabling it. When calculating the cost of introducing a new service, migration from the old one should be factored in. When looking at replacing storage and migrating the data, how long will it take to move the data, or will the system be down when it is happening?

Without innovation customers will begin to look outside of the ITO for services. For example, say an ITO offered backup as a service ten years ago. Back then, there was no 'hot' database backup; applications had to be shut down before the backup could be done. Technical innovation enabled hot database backup; if the ITO did not embrace this into a new service, then the customers would probably have looked elsewhere or even done this themselves in order to get the business benefit of not having to shut their applications down just to get a backup. Similarly, if customers believe (and perception is everything) that they can get a service cheaper outside the ITO, then they will. It is up to the ITO to prove that the services it provides are not only technologically advanced, but also good value for money.

---

[1] Serial ATA is the next generation of ATA interface and, because of the improved speeds and comparative cheapness of the disks compared to SCSI, a new possibility for many enterprise applications.

Innovation is, therefore, a double-edged sword; without it, the customers will desert in favor of new suppliers, but too much, and the ITO will revert to chaos as every technological advance becomes a service for a single customer.

> There is only one thing worse than speaking to no customers, and that is only speaking to one.

The only way for the ITO of the future to be successful is to have a continuous two-way discussion with *all* its customers. With their input and help it will be possible to design new services that are useful to all. Continuous improvement to process and increased automation will further increase efficiency and drive down costs, ensuring IT delivers real business value – and can prove it.

14.4 SUMMARY

- Within a utility computing environment, creating one-off services for individual customers will result in a return to the chaos and inefficiency that exists in the ITO today.
- Continuing to do business as today is no longer an option for the ITO of tomorrow; a more flexible environment is required to make best use of the company's assets.
- Your customers are your reason to exist. Talk to them, understand their needs and provide services they need.
- Innovative new services based on new technologies must actively be found for the ITO, and ultimately the business, to remain competitive.

Appendix A

Case Studies

Much of this book has focused on theory, methodology and best practice for utility computing transformational initiatives. Hopefully, this reference material will prove to be invaluable to IT strategists who want to drive the utility computing concept into their organization. However, the authors recognize that the best way to illustrate the power of a concept is to show how it has been adopted in real life, and how true business value was derived as a result.

This section provides an overview of several organizations that have benefitted from the utility computing approach, and that have utilized successfully some, or all, of the methodologies and best practices that are referenced within this book. The reader will notice that, in line with the principles discussed earlier in the book, the clients that are featured in this section have, without exception, all approached utility computing in a focused and pragmatic way that addresses specific business issues such as driving down cost or improving time-to-market while dealing with higher expectations from their customers (see Figure A.1). None of the clients described here have created a 'data center utility' (i.e. transformed their entire IT estate to operate in this way). In reality, successful implementations of the utility model have involved a specific scope for transformation (in the case of our examples, specific scopes include a disaster recovery solution, an enterprise data backup and archiving solution and an online enterprise storage solution). The last case study is, in part, hypothetical; there are a number of customers who are putting server or application utilities in place, but

Delivering Utility Computing.   Guy Bunker and Darren Thomson
© 2006 VERITAS Software Corporation. All rights reserved.

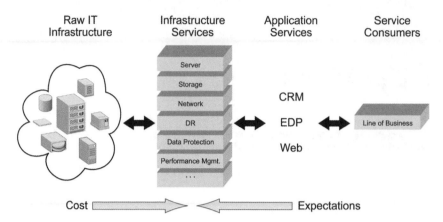

Figure A.1 Delivering tightly scoped IT utilities to the business.

these are incomplete at the time of going to press. Rather than leave an example of this out, this case study has been written up as far as implemented and then extrapolated.

A.1 CASE STUDY: DARTMOUTH COLLEGE'S CAMPUS-WIDE BACKUP UTILITY[1]

A.1.1 Overview

Dartmouth College is a private, four-year, coeducational undergraduate college with graduate schools of business, engineering and medicine and graduate programs specializing in the arts and sciences. Dartmouth is the ninth-oldest college in America, founded in 1769.

With over 30 separate departments on its distributed campus (see Figure A.2), each with their own ideas as to how IT can best serve them, Dartmouth had been thinking for some time about ways in which it could standardize IT services and then provide these to its various departments using a centralized 'service provider' model.

Early in 2004, Dartmouth elected to move towards a utility computing model within IT, and decided that its storage environment (and specifically its system backup service for unstructured data) would be a good starting point for utility transformation.

---

[1] Case study reproduced by permission of Dartmouth College.

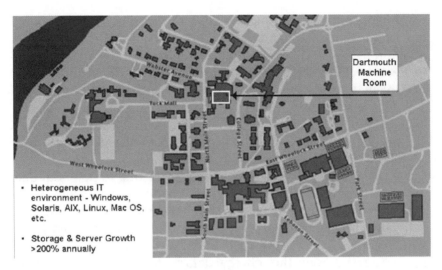

Dartmouth
Machine
Room

• Heterogeneous IT
  environment - Windows,
  Solaris, AIX, Linux, Mac OS,
  etc.

• Storage & Server Growth
  >200% annually

Figure A.2 The Dartmouth College site in Hanover, New Hampshire, USA.
Reproduced by permission of Dartmouth College.

A.1.2   Business challenges

There were several major business challenges that the utility computing project at Dartmouth was designed to help resolve. These
challenges included:

- *Complexity within the data center* including disparate standards in
  storage and server design, complex designs leading to high cost
  of management and a very complex data network.
- *Underutilized IT assets* and limited visibility into ongoing capacity
  requirements.
- *A need to improve the connection between the business requirements
  and IT services* in terms of data retention and protection.

These challenges were actually apparent across most of the IT
services at Dartmouth, but the IT organization recognized early on
that 'boiling the ocean' and trying to transform too much with a
new model like utility computing was probably not sensible. The
issues listed above were particularly pertinent to the data protection IT service, and this was soon chosen as the target for utility
transformation.

During the initial analysis of the current data backup infrastructure at Dartmouth, they discovered that, although system backup

would appear to be a simple and basic service that could be offered in a uniform manner across all departments, this was not the case at all. In fact, a study found that the methods and standards used to create backups across the multitude of college departments were massively diverse.

Some departments were very happy with their 'home-grown' backup solutions, even though they were owned and administered by non-IT personnel. Others had actually not felt compelled to back up their data at all!

All of this highlighted significant risk (relating to potential data loss) and inefficiency (involving high cost of management of the very disparate and non-standardized backup environments).

In addition, the users of IT at Dartmouth had very mixed views with regard to how good a job the IT department was doing in protecting data. Some thought that data protection was just 'overkill' in their case, whilst others were very concerned that their data was not as well protected as it could have been.

Enter the utility computing model...

A.1.3 Solution overview

Once data protection had been targeted for utility transformation, a design initiative was undertaken to establish certain service characteristics, including:

- What backup services were actually required by IT users?
- How would these services be offered and presented to the various college departments?
- What organizational structure would need to be adopted to support the utility backup service?
- What technical architectures would be used to support the initiative?

Readers will notice that these fundamental characteristics can be captured through use of the Utility Computing Reference Model (discussed at length in Part Two of this book).

IT executives decided to roll out a pilot of the service based on key issues that existed within the college, including the backup of college departmental records, files and structured data such as databases. It was decided that, by centralizing backups via a utility service, IT

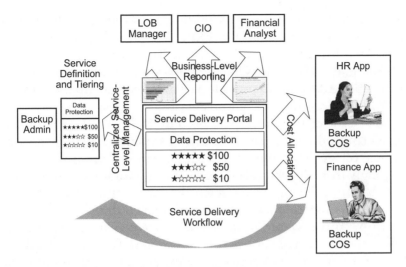

Figure A.3 The Data Protection Utility at work.

staff would then be able to focus on more strategic and consultative issues instead of the complex and very manually intensive tasks associated with the traditional system backup method.

Encapsulated within the planning of objectives for the new utility were the following design goals:

- *Centralize IT* – by standardizing backup policy and standards whilst, at the same time, creating a centralized backup infrastructure and moving to implement a storage area network.

- *Improve efficiency* – by monitoring closely the growth of backup volumes over time so as to predict accurately required storage capacity and drive up utilization rates.

- *Improve IT/business relations* – by giving IT executives the ability to align data protection services with real business requirements and to measure the success of this over time.

Launched in 2004, the Data Protection Utility at Dartmouth (Figure A.3) has so far been rolled out to 500 users across 37 college departments. The plan is to roll it out to 2000 users this coming fiscal year, and an additional 2000 users in the following fiscal year.

Having designed and documented the attributes of the service classes and created service level expectations (SLEs)[2],

---

[2] As opposed to the more stringent service level agreement, which often implies financial penalties and creates undue stress for all involved when initially setting up an IT service.

organizational structure and workflows associated with the new utility, Dartmouth selected the VERITAS software portfolio as the foundation for the new utility service. This included using VERI-TAS Foundation Suite (for Oracle database and application servers), VERITAS NetBackup (to provide a centralized backup mechanism) and VERITAS CommandCentral Service.

Some of the challenges experienced in deploying the solution involved controlling the consumption of storage capacity across the various college departments. In addition, consumption already varied greatly from department to department with some teams choosing to back up all of their data (regardless of its criticality) and others not backing up data at all. Using the utility computing model, Dartmouth executives were able to establish a creative financing strategy (based on the class of service model, see Part Two of the book) effectively to minimize overall capital expenditure. Dartmouth now charges users for capacity used at the end of each month and the costs associated.

Quality of backups (relating to factors such as performance, data retention/data protection method and service availability) were made variable (based on a class of service model) and presented to IT users using a rate card (see Figure A.4). Users could then choose which class of backup service would best suit their department's specific data protection requirements. Initially, users were educated on their usage levels and the costs associated with these. They were then given a month's grace period to reassess what they were backing up. This had an enormous overall impact right away on the amount (and type) of data that would be backed up by the utility service moving forward. The college based a 'fee per GB' pricing on estimates that calculated the leasing costs for hardware, software costs and labor expenditure for offering each class of backup service. Finally, by keeping close track of storage capacity, IT staff could now have accurate and up-to-date usage reporting (by department), as well as clarity on how fast this capacity was growing month on month.

A.1.4   Business benefits

From a financial perspective, once fully deployed and operational, the Data Protection Utility had a goal to generate enough revenue (albeit internal revenue) to pay for the project fully within the first

Backup Class	Server	Desktop
Full Backup frequency	Weekly	Monthly
Incremental backup free-quency	Daily	Daily
Data Retention time	Four weeks for weekly back-ups and then a year for the monthly backups.	90 days

Figure A.4 A Data Protection Utility rate card.

fiscal year. Additionally, the true value of protecting departmental data could be understood better by IT users (who had to think seriously about their own data protection requirements).

Other business benefits delivered by the project included:

- *Headcount savings*. A centralized system required far less technical staff than a disparate, non-standardized system.
- *Physical resource savings*. An enormous amount of resource savings were made as a result of no longer having IT systems for backup housed by each of the 37 college departments. Additionally, a centralized system could be housed in a well-controlled and physically suitable data center environment.
- *Service improvement*. As a result of taking a collaborative and consultative approach to the project, IT and business users worked together to decide what the utility data protection service offerings should look like. This brought overall service improvement and alignment between IT and the business.
- *The ability to showcase the utility computing model*. Dartmouth intends to use the model to transform many other areas of the IT estate. This will now be easier to fund and sell to the business due to the data protection project being perceived as a success.

A.1.5 Lessons learned

Of course, nobody's first utility computing project is going to run without a hitch. In analyzing how they could have improved the roll out of the Data Protection Utility project, Dartmouth concluded that several lessons could be learned and used to improve future utility transformations at the college.

First, the project experienced some confusion between the college departments and the IT group. These were caused mainly by a lack of understanding (on the business's part) of IT terminology and of utility computing principles. Within a service-driven environment, it is important that the IT group can converse with the business, and resolving terminology issues is important to do upfront. New technology, coupled with new principles, creates new terminology, which can be misinterpreted. In hindsight, it would have been a good idea to deliver a series of educational workshops for the consumers of the utility. These workshops would have provided an excellent platform for the discussion and debate of project principles, goals and terminology.

Looking back at their project, Dartmouth IT executives also concluded that the project would have run smoother if there had been more frequent interactions between the IT group and the business sponsors (departmental heads, etc.). The project should have scheduled formal meetings between IT and the business every month to discuss potential issues. Dartmouth also concluded that these meetings should probably have occurred on a weekly basis at the start of the project. This would have improved project communication and helped to avoid some of the cultural issues that can arise in a project of this type.

Dartmouth IT also concluded that they should have utilized the capabilities and personnel outside the IT group to help in the design, introduction and support of the Data Protection Utility. Specifically, the IT helpdesk, education and first-level technical support, would have increased the buy-in and alleviated the initial increased workload that occurred when the utility model was in its early stage of adoption.

A.1.6 Next steps

The Data Protection Utility is seen as a huge success by IT and business executives alike. Dartmouth is now examining how to use the utility computing model to transform other IT infrastructure services, such as online storage and server provisioning.

There are plans to get a flat rate for billing based on usage. Much like a cell phone bill that has a fixed number of minutes for a fixed price, the plan is for a flat rate for the first X GB backed up. An additional bill would be created for any data backed up above the limit agreed upon as part of the SLE, but it would be the exception rather than the rule. This will help the different groups in their planning as it introduces a 'fixed' cost, rather than a fluctuating one.

Finally, a detailed return on investment (ROI) study is underway to prove the value of moving to a utility computing model.

A.2   CASE STUDY: DIGITAL TV CO'S DISASTER RECOVERY UTILITY

A.2.1   Overview

For legal reasons, we were not able to provide the name of the organization that features in our next case study. We will refer to it as 'Digital TV Co.' (DTC) in this case study.

DTC is the largest supplier of digital television services in the UK. It provides broadcasting of sports, movies, entertainment and news programs to around seven million households across the UK. The organization is also instrumental in changing the face of entertainment in the United Kingdom. More than 17 million viewers in seven million UK households enjoy an unprecedented choice of movies, news, entertainment and sports channels and interactive services provided by DTC.

Clearly, a business of this type is very heavily reliant on leading-edge technology (and rigorous governance and process to surround its technology assets). Prior to the project described in this section, the organization operated in excess of 15 data centers, housing IT systems that supported many different facets of the business, including program scheduling, customer billing, website support, encryption, customer relationship management systems and broadcast platforms.

A.2.2   Business challenges

The huge success of DTC's digital television service offerings during the past seven years has led to several IT-related business challenges. In general terms, the demand for IT capacity over time can be fairly unpredictable (they are constantly introducing new digital services to the market, all of which demand more IT resources). Additionally, 'time to market' is a critical success factor in introducing these services, since DTC is now facing an increased amount of competition in its market.

Other IT-related business challenges include:

- *A huge amount of physically dispersed IT systems* spanning many separate data centers, some large and some very small.

- *Complexity* – as an aggressive adopter of leading-edge technology, distributed systems have become more and more difficult to govern and manage.

- *Business risk in the case of site disaster* – most of DTC's existing data centers are actually sited at the same location in London.

- *Lack of technology standards* – the huge explosion in IT asset had left the IT organization with a plethora of technologies and a lack of standard best practices for providing IT services visibility into ongoing capacity requirements.

- *A lack of standardized IT services to offer the business* causing 'time to provision' issues that impact the 'time to market' profiles of the consumer offerings.

In order to address these challenges, a new IT project named DCC (Data Centre Consolidation) was initiated in 2003. Whilst the project's key objectives were to drive standardization and consolidation, it also sought to provide solutions to the other business issues listed above.

During the initial strategic discussion that led to the planning of the DCC project, IT executives and strategists within DTC started to research and investigate utility computing theory (much of which is discussed within this book). Strategists at DTC became particularly interested in the class of service model and how this approach to providing standardized IT services could help the organization to address some of the critical issues, described above.

A.2.3 Solution overview

Initial utility computing workshops were run at DTC to ascertain how the principles inherent within the utility model could help the DCC project. After some discussion, IT executives concluded that

Applications	Business criticality
TV transmission support systems	Mission critical
Supply chain management, Email	Business critical
Enterprise back office systems	Business important
Departmental functions	Function important
Standalone systems	Non-critical

Table A.1   Application criticality at DTC

the IT service type that the business was most focused on (in the context of what the DCC project would provide) was disaster recovery. The business was conscious that, should disaster occur at DTC's headquarters in London, this could render the organization unable to provide a service to its subscribers.

Work was then carried out to ascertain what a 'disaster recovery (DR) utility' would look like (once again, not from a technology perspective but from the perspective of the consumer of this IT service – the outside layer of the Utility Computing Reference Model). The following data points needed to be captured by the IT organization before attempting to design the new utility service:

- What DR services would be required by the organization's business lines?

- What would the service level agreements representing these service have in them?

- How many classes of service would be needed to cover all DR requirements?

Through a process of collaboration with the IT users and various departments at DTC, the IT group concluded that the business ran applications that could be categorized into five levels of application criticality. (See Table A.1)

Once DTC had discovered that all of its applications could be categorized in this way, it could start to think about building a DR service that contained a class of service model that mapped to the five application types. Once again, a consultative and collaborative process was undertaken to canvas opinion, discuss and debate what quality of DR service would be relevant to the five application types. At this point, IT staff started to refer to the five classes of service as 'the metal standards' (platinum, gold, silver, bronze and copper). Much conversations and debate then occurred between the

Service class	Example application	Service levels
Platinum (Immediate recovery)	TV transmission support systems	24 × 7 scheduled 99.99% availability RTO=2 hrs, RPO=0 hrs
Gold (Hot standby, redundant)	Supply chain management, email	24 × 6 3/4 scheduled 99.5% availability RTO=8 hrs, RPO=4 hrs
Silver (Hot standby, reuse)	Enterprise back office systems	18 × 7 scheduled 99.0% availability RTO=3 days, RPO=1 day
Bronze (Intermediate recovery)	Departmental functions	18 × 7 scheduled 98.0% availability RTO=5 days, RPO=1 day
Copper (Gradual recovery)	Standalone systems	12 × 5 scheduled 98.0% availability RTO=10 days, RPO=1 day

Table A.2 Disaster recovery service level objectives

IT group, business leaders, application owners and IT governance staff within the organization. Through this consultative process, the IT group was able eventually to map some basic service level objectives to the five service classes (See Table A.2).

The 'RPO' and 'RTO' service level characteristics mentioned in Table A.2 stand for 'recovery point objective' (how much data can the business afford to lose subsequent to a disaster?) and 'recovery time objective' (how quickly should an application be operational again after a disaster has occurred?). Next came the task of analyzing each of the 50 or so application services at DTC so as to ascertain which DR service class would be used by which applications within the business. Of course, most application owners stated initially that their application was very critical (a very subjective term!) and would, therefore, require the 'Platinum' level of DR service. Only through a formalized process of 'scoring' each application, based on its true characteristics, were DTC able to decide how the various applications within the business should map to the five classes of service.

A 'scorecard' was developed that could then be used to rate each application, based on the following characteristics:

- financial impact of application downtime;
- user expectation of application availability;

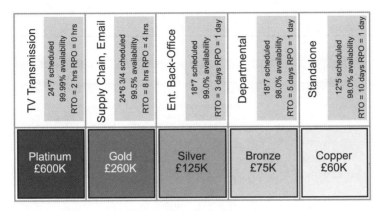

Figure A.5 Estimated service delivery costs.

- usage profiles;
- application failure history;
- brand impact caused by application outage.

Surprisingly, once this rating exercise had been carried out, DTC found that very few applications actually required a 'Platinum' grade of DR service. The majority of the applications in the business required a 'Gold' or 'Silver' service, and around 15 of the applications were actually used for test and development work and only required a minimum class of service ('Copper'). These findings were important; they meant that, for the first time, the IT group could provide an IT service that they knew was appropriate and relevant to the business. It also meant that a huge amount of cost could be saved (since the delivery of a 'Platinum' service cost a lot more than a 'Gold' one, See Figure A.5).

At this stage (and for the first time) the IT group could start to develop architectural standards ('blueprints') to support each of the classes of service that would need to be provided by the DR utility. Architects could now design infrastructures that were absolutely appropriate and relevant to business need, and blueprints could be authored that ensured that IT infrastructure design would follow the same best practices every time (i.e. there would be a 'recipe' that could be followed for building a 'Gold' IT infrastructure). In addition, these technical blueprints could stipulate standards that should be adhered to whilst designing new infrastructure (see Section A.2.2).

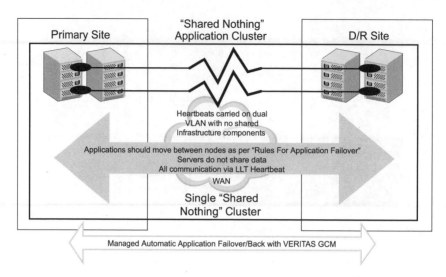

Figure A.6 Screenshot from 'architectural blueprint' tool.

A tool was developed that allowed IT designers and engineers to refer to the new architectural blueprints to ensure that their new design were compliant with IT policy (Figure A.6). This tool included approved technology standards, approved vendors, best practices in design, technical schematics and reference to IT governance policy. IT personnel wishing to create new infrastructure within the new data centers at DTC that were not compliant with the new standards would need to obtain approval at the IT Director level.

Whilst all of this revised IT strategy and design effort was occurring (the effort described within this section took around six months), DTC had been building new data center premises to house the newly consolidated IT environments. Once these new data centers were operational, DTC could start to deploy new standardized systems in the knowledge that many of their original project objectives had been met.

A.2.4 Business benefits

The utility computing approach to IT/business alignment and technology design brought significant benefit to the DTC IT operation. If we refer back to the project objectives, detailed earlier, we can see just how effective the approach was:

- *A huge amount of physically dispersed IT systems* – all IT systems are now being migrated to two new data centers. The standardization and class-of-service-driven approach to IT design is driving a huge amount of physical consolidation into the IT estate at DTC.

- *Complexity* – creating standard design blueprints for infrastructure has allowed DTC vastly to reduce the number of suppliers that it deals with. In addition, all system designs are now based on standard templates, making life a lot less complicated.

- *Business risk in the case of site disaster* – not only are all applications now protected from the effects of a site disaster, but their level of protection is based on a well-understood class of service model and application owners can be sure of what to expect in the case of disaster.

- *Lack of technology standards* – new IT systems that are deployed by the DCC project are now standardized completely and based on the architectural blueprints that have been authored.

- *A lack of standardized IT services to offer the business* – the business can now request standard DR services ('the metal standards'). This has brought effective alignment between the IT group and the business in the context of DR.

A.2.5   Lessons learned

Above all, DTC learned a valuable lesson concerning the importance of collaboration during this project.

The advent of a class of service model for something as critical to DTC as disaster recovery would not have worked had it not been for IT executives and strategists ensuring that their consumers (the business lines) were involved in most of the decision making that occurred through the lifecycle of the DCC project. Business representatives were involved in conversations ranging from 'how many classes of service do you need?' to 'what should the SLA contain?' and 'what revenue does your application generate each year?' The IT group at DTC feels that its relationships and understanding concerning the business have been improved as a result of this project approach.

Another lesson learned during the project was, once again, (see the previous case study) concerning the use of IT language. Some confusion occurred at DTC as business people became aware of the

DR classes of service model. For DTC to provide its service to its subscribers, many layers of technology (from storage systems to satellite dishes and set-top boxes) have to be leveraged. The scope of the DCC project was to ensure that IT infrastructure (supporting business applications) was able to recover from a site disaster. This, of course, did not mean that in the event of all types of disaster subscribers would be guaranteed no loss of service. The solution described here would not, for example, protect from loss of service in the event of satellite failure! The lesson: be very clear when talking to non-IT business executives about the scope of what your utility will provide.

A.2.6 Next steps

At the time of this book going to press, the DCC project team at DTC was in the process of migrating its historic IT infrastructure to the new, standardized and consolidated design. Early signs are that the business benefits described earlier would all be realized fully as the project neared completion.

IT executives at DTC are keen to explore how the utility computing model could help them to reduce cost in others areas of IT (such as storage).

A.3 CASE STUDY: ARSENAL DIGITAL SOLUTIONS' INFORMATION STORAGE UTILITY[3]

A.3.1 Overview

This case study differs slightly from the previous two, since it describes the utility computing method being used by a business in order to provide utility infrastructure services externally to remote customers.

Founded in 1998, Arsenal Digital Solutions centrally manages one of the largest multivendor storage utility environments in the world. With more than 10 Petabytes of managed client data and 30 data

---

[3] Case study reproduced with permission of Arsenal Digital Solutions.

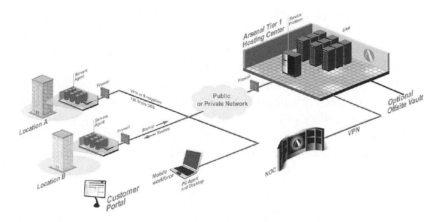

Figure A.7 Remote data protection utility, 'ViaRemote'.

centers spanning five continents, the company's storage services are used by over 900 businesses worldwide, including many Fortune 500 companies. From its centralized and automated utility infrastructure (Figure A.7), Arsenal delivers a comprehensive portfolio of fully managed, on-demand solutions that support the complex storage, data protection, business continuity, disaster recovery and compliance needs of its customers.

As businesses grow, the demand for IT expertise and reliability also grows. Clients today want to realize the benefits of IT investment and require rapid, reliable and consistent delivery of IT service. By providing centrally managed, turn-key, utility-based IT services, Arsenal enables IT organizations to reallocate their own resources to support core business activities, rather than the complexities of infrastructure. When Arsenal's clients understand the benefits of data storage reliability, they begin to derive the benefits of the utility computing model and shift away from the time-consuming process of maintaining and managing rapidly evolving technology.

A.3.2 Business challenges

The provisioning of utility-based storage services has proven to be hugely successful for Arsenal and the company continues to innovate, bring new utility services to market and grow. As an example, due to increased demand for its data protection, business continuity and regulatory compliance services, in 2004, Arsenal performed 2.5 million data backups and 3500 restores.

Arsenal's success as a business (and the success of its clients) relies on a centrally managed storage infrastructure and the implementation of utility computing best practices. Without a carefully balanced combination of the principles discussed in this book, Arsenal would find it difficult to address the following critical business issues:

- *Flexibility.* Arsenal requires the ability to provide standard utility services effectively to organizations ranging from small and medium businesses to large enterprises.

- *Availability.* Clients simply expect utility services always to be available and monitored 24 × 7 with expert staffing.

- *Reliability.* With new US federal regulations requiring that key industries incorporate out-of-region backup facilities for data protection and operational support, the reliability of a storage utility service to these clients is clearly of critical importance.

- *Scalability.* Future business success at Arsenal will be dependent largely on its ability to scale the infrastructure and deliver new, innovative services designed to address data growth, estimated at over 80% per year.

In order to address these challenges, Arsenal became an early adopter of utility computing. Its business has become very successful through its focus, not only on the technology components that enable this strategy, but also on rigorous attention to service definition, process design, standardization, automation and a constant drive for maximum operational efficiency.

Without operational efficiency (derived through utility computing principles), a business such as Arsenal's would suffer from one of two possible dynamics. They would either lose money because of overspending to ensure that their client's SLAs are met, or they would have to price their services to overcome the extra internal expenditure necessary to run an inefficient, non-standardized infrastructure.

Arsenal Digital Solutions is a perfect example of how utility computing can be made to work by giving focus to the disciplines described within this book.

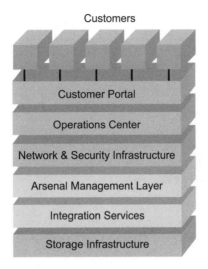

Customers

Figure A.8   Arsenal's s3 Service Architecture™.

A.3.3   Solution overview

Arsenal has developed and implemented a utility computing method and platform to deliver high-quality, repeatable service offerings. Arsenal's s3 Service Architecture™ (incorporating both process and architectural elements) is the key enabling platform behind Arsenal's storage management services (Figure A.8). This patent-pending architecture combines proven technology and formalized processes and procedures with unique intellectual property, enabling Arsenal to deliver the most reliable architecture for deploying, managing and supporting its storage management services. It enables Arsenal's customers to minimize their dependence on complex internal storage architectures, point solutions or specially trained staff with the confidence that Arsenal has their data protected, secure and accessible.

Why is the s3 Service Architecture™ so important? Because it provides the foundation for Arsenal to deliver financially guaranteed service levels consistently and exceptional customer satisfaction, regardless of what the service is.

As you would expect, Arsenal's approach to providing utility computing services involves much more than hardware and software technology components. It is a complete outsourced storage management service solution designed to improve operational

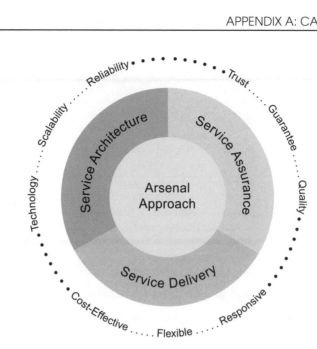

Figure A.9 The Arsenal approach to delivery utility services.

effectiveness, increase quality and consistency of service and simplify the management of complex enterprise storage environments.

The Arsenal approach (Figure A.9) combines three key components to create compelling and reliable utility-based storage services:

- service architecture (s3™);
- service delivery;
- service assurance.

Through the development of appropriate service management standards and architectural blueprints, Arsenal now provides clients with a whole range of data protection, storage management and business continuity services. These services are fully productized and standardized so that clients find it very compelling to subscribe to them. They are also marketed very professionally to maximize market potential (see Section 7.2).

A.3.4   Business benefits

The utility computing approach to developing Arsenal's outsourcing business fundamentally allowed Arsenal to differentiate itself from other, similar, organizations that were providing traditional outsourcing services within the US. By adopting utility computing, Arsenal has managed to create a value proposition and range of service offerings that differentiate it in the following ways:

- *Clients can dramatically reduce total cost of ownership (TCO)* – Arsenal's shared infrastructure is run more efficiently than a typical business operation and, as a result, clients can save money by leveraging Arsenal's best practices and architectures.
- *Clients can expect a flexible, scalable model for infrastructure operation* – the utility computing approach allows clients to scale up and down dynamically, based on business requirements over time.
- *Arsenal's pricing is extremely competitive* – it can afford to aggressively price its offerings due to its extremely efficient mode of operation and world-class infrastructure.
- *Arsenal is now an expert in utility computing* – this attracts new clients, who tend to be very compelled by the huge success. that Arsenal has already seen and by the high levels of customer satisfaction that they can reference with existing clients.

A.3.5 Lessons learned

Arsenal learned that becoming an early adopter of a new approach to doing things is not necessarily a bad thing. Many organizations elect not to adopt utility computing since they see the approach as potentially creating too much business risk and disruption in day-to-day operations.

Our experience (and Arsenal's) is that this approach to the transformation of IT services will bring significant business benefit if approached pragmatically and with a few basic guidelines in mind:

- Talk to your potential customers (at length) about what they really require before embarking on your utility computing initiative.
- Make your utility service compelling and differentiated.
- Do not try to transform services that do not naturally lend themselves to the utility model.
- Focus initially on service definition, then on process and best practice and finally on technology.

- Create 'pilots' to test the approach before betting your business on the model.
- Continue to innovate and build services that enhance the strategic value of the utility service and strengthen the relationship with customers.

A.3.6 Next steps

As has already been mentioned, Arsenal has seen enormous business success in providing its utility storage services to clients in the US, Asia, Europe and Latin America. As the constrained capital market forces IT executives to look beyond the traditional boundary of their in-house organization, Arsenal is well positioned to offer customers comprehensive solutions designed to meet today's and tomorrow's strategic business imperatives.

A.4   CASE STUDY: A TELECOMMUNICATIONS SERVER AND APPLICATION UTILITY

A.4.1   Overview

At the time of going to press, there is not a completed installation of a server utility. However, several companies are going through the implementation and, in order to cover this important utility, a faux case study for a company called TelcoZ is presented.

Telecommunications companies continue to be at the cutting edge of IT environments and implementations; however, the increasing competition has led to cost reduction drives as well as requirements to become more flexible when it comes to deploying new services. Often, the services only have a life span of a few months before the competitive advantage is lost. The introduction of utility computing principles and processes has enabled TelcoZ to respond to the changing business requirements.

TelcoZ currently has more than 10 000 servers, many of which are in different countries across the world. The company is continuously going through an upgrade program, but is now embarking on a complete technology replacement program, which is scheduled to take three years.

A.4.2 Business challenges

Current server utilization is running at less than 20% and the CIO has highlighted the need for improvement in order to justify the technology replacement program. New business ideas are often delayed, as the response time of the ITO is slow when compared to the time a service is likely to remain a competitive advantage.

The ability to provision new applications rapidly and deliver the corresponding infrastructure to the business in as short a timeframe as possible is key to the success of the utility program. Server utilization also needs to be improved and, while much of that will come from consolidation, it is also imperative that the environment remains flexible.

TelcoZ's success moving forward will depend on addressing the following business-critical issues:

- *Time to service delivery*. Reduce time from service request to delivery to hours rather than weeks.
- *Flexibility*. The ability to deliver both small and large application implementations 100 GB through to 1 TB.
- *Availability/reliability*. Lines of business expect their service to be always available.
- *Scalability*. Not just for the infrastructure itself, but also in the service request and delivery mechanism.

TelcoZ adopted utility computing as much for the process as for the technology. The technology refresh gives, in effect, a green field project, which pushes standardization and, therefore, makes automation simpler. It also meant that the *asset ownership* problem was reduced; this has also been helped by a push from the CIO that all assets will be owned and managed by the ITO, rather than the individual lines of business.

Applications would be monitored through service level agreements, which would be based primarily on end user response time and number of simultaneous users. Accounting for the service would be provided in the form of a chargeback report, but there would not be any real chargeback, instead this would be used as a visibility into the costs. The costs would be calculated at the beginning of each 12-month period and would remain constant throughout the year. Costs would include hardware, software and personnel, but would not be split into any great detail.

A.4.3 Solution overview

TelcoZ is part-way through its utility computing initiative; time has been spent developing the service templates that are to be used as the standards for the application platforms. By creating template solutions, the company has been able to negotiate supplier deals for delivery of new hardware, reducing the order time down from weeks to days. However, its preferred mode of operation is to utilize spare capacity in the environment. Currently targeted to development environments, the possibility of sharing resources is increased greatly and it is easier to drive up the utilization as applications can be installed on existing hardware.

The approach that TelcoZ took was more than just a change in hardware and software, it also involved the start of some cultural change. Breaking out of the asset ownership issues is not easy, and while CIO-level backing has been successful in the development environment, it remains to be seen how easy it will be in the production environment.

The development of the application and server templates alone would not have given all the benefits seen. Time was spent developing a number of process automation workflows that enable automation of the most common tasks:

- provisioning a new server + storage (including adding it to the backup policies);
- increasing (and decreasing) storage;
- installing and configuring specific applications (database, J2EE application server and web server);
- deprovisioning a server.

The development of the workflows has enabled operators at TelcoZ to carry out tasks that were previously only carried out by administrators. While comparatively simple with what is planned, they have already enabled both scalability and responsiveness and will act as a good base for the future.

A.4.4 Business benefits

TelcoZ has benefitted in a number of ways already:

- *Users understand their usage and costs associated with it* – the reporting built into the environment means that usage reports can be created on a regular basis and the various users kept informed. By associating costs with the usage, the users can then determine themselves whether this is good usage of budget. Flexibility in the infrastructure makes it relatively simple for them to request that applications be moved to different levels of service.

- *Reliable response to requests* – one of the biggest benefits has been that the users understand what they will get and when they will get it. This is not only for initial requests for new application instances and servers, but also for changes in existing services, whether it is increasing the storage or moving the application to a bigger server.

- *Time for innovation* – while the template environment initially seems dull for the administrator, it has resulted in more time to investigate new technologies to incorporate into new services. Much of this was done upfront when initially designing the services, however, it will continue so that new or replacement services can be introduced successfully .

- *Decreased redundant resources* – resources that have been used for projects that have not made it to production are returned to the ITO for reuse rather than sitting idle.

A.4.5 Lessons learned

TelcoZ learned that while it is easy to develop templates with the help of their users, it is all too easy for those same customers to ask that they be excused from using them, i.e. that they are a special case and so should be exempt from using the templated solution! A strong will was needed to overcome this barrier, and it is probably an ongoing battle.

There was a great temptation from within the ITO to provide custom solutions, and it was seen as a conflict of interests not to. On one hand, the ITO was supposed to be a customer-driven organization ('the customer is always right') and on the other, an efficient organization, in which case the customer does not necessarily get what they want. It was fortunate that the project was a green field one, although this also meant that the normal 'day job' had to continue, and so there was often the request to 'just do it as you did before'.

The key take-aways were:

- Get top-level executive sponsorship, for example the CIO, CFO or CEO, before you start. Ensure that they are seen to be driving the project as it will help break the asset-ownership problems.
- Talk to the users in order to create the template solutions. This is not a one-off, but needs iterations. A trial installation of an application will probably be needed to prove that it does work.
- Expect all your users to think of themselves as 'special cases' and ensure that you have a process for dealing with them. This will undoubtedly involve escalation to the CIO, hence the need for executive sponsorship. A 'special case' is a failure of the utility computing model and will result in chaos.
- Spend time on process. Now is a good time to review current processes and to look for improvements, use of process workflow technology can help in standardization.
- It is essential to create only a handful of service levels and ensure that users categorize their applications into them. A service with only one user is a one-off and (comparatively) expensive to run.

A.4.6 Next steps

Initial trials have proved to be successful both in terms of ROI and popularity with the business units, in being able to respond to new business opportunities more rapidly. While the utilization in the development environment has been increased from less than 20% to more than 50%, there is still the possibility of increasing it further. The other challenge is going to be moving it into the production environment.

Virtual machines are being explored on two fronts, the first being to insulate applications from each other, including in the production environment. The second being a way to provision servers even more rapidly; by deploying a virtual machine as a part of a server, it is hoped to automate the request for a server fully and provision it without any administrator intervention.

Appendix B

Utility Computing Planning Forms

These forms are examples; the chances are that they will have to be modified a little before they can be used by IT departments. Careful form design is important; all required information should be collected in one pass. Continually returning to customers to request more information will not make the IT utility popular.

In many cases, the answer to a question will be 'no' or 'not known', these are just as valid as any other. Remember the data collected here is to enable you to put together a plan to move from the traditional data center approach of running IT, to the more flexible and efficient way that is utility computing.

B.1 BASELINING

Application name	Department	Business criticality (1–5)	Availability (1–5)	Standard services used	Metered usage	Downtime cost per hour
e.g. SAP R3	Finance	1	1	Backup, storage provisioning, server provisioning, client provisioning and roll out	No	$2M

Delivering Utility Computing. Guy Bunker and Darren Thomson

Business criticality terms should be defined specifically for each enterprise, e.g.:

1 – Critical, e.g. online shop, CRM.

4 – Not critical, e.g. payroll.

5 – Redundant, reporting that was useful five years ago.

Availability terms should be defined specifically for each enterprise, e.g.:

1 – Highest, e.g. replicated data, wide area failover, clustered, tape backup.

4 – Low, backup once a week.

5 – Lowest, not even backed up.

B.2 BASELINING 2

Application name	Operating system	Storage usage	CPU usage (if known)	Network usage (if known)	Backup copies	Clustered	Replication
e.g. SAP R3	Unix, version X, patch Y	200 GB			Full weekly Incremental daily Monthly copies held for three years	Yes. Locally and globally	Yes. Mirrored storage locally, wide area copy

B.3 USER DEPARTMENT VIEW

		Department		
IT service required	Dependent	Taken for granted or planned?	Current implementation: enabler or inhibitor	Deficiencies
e.g. backup	Yes	Taken for granted	Enabler	Slow to add new servers

B.4   IT VIEW

Service	Deptartmants serviced	OSs supported	One-off?	Financial metrics?	Performance metrics?	Accountable?	Chargeback?
e.g., backup	All	All (Windows, Solaris, HP/UX, AIX, Linux, etc.)	No	No	Yes (limited)	No	No
e.g. DB2	Finance	AIX	Yes	No	No	No	No

B.5   IT TECHNOLOGY SUPPORT

Department	# Servers	# OSs used	# Storage/ # storage types	# Applications/ # Application types	# Networks
e.g. Finance	20	three AIX, ten Solaris, seven Windows	Not clear, some EMC, some IBM, some Sun	one SAP, three Oracle, one DB2, ...	two CISCO

B.6   IT PLANNING

Service	Score	Utility/ factory/ consultancy/ operation	Availability expectations	Performance expectations	Growth expectations	IT cost Growth	Organizational structure
e.g. backup	1	Utility	24×7	SLA/ backup window restrictions	50% year on year	???	Centralized management by IT department, local users enabled for restore

B.7   COST SAVINGS

Hardware	Software	Owner/ primary user	Original cost	Maintenance/ support cost per year	End of life/ Reprovision
E.g. Server X		Finance	10000	2000	Y (now obsolete)
	Application A	Engineering	2000	0	Y (replaced with newer version, no relevant data stored)

B.8   HARD METRICS

Hardware	Original cost	Initial utilization	Final utilization	% cost saved

	Before UC initiative	After UC initiative
# Administrators		
GB storage		
GB/Administrator		
Time to provision storage		

	Before UC initiative	After UC initiative
Online data copies		
Offline data copies		
Average offline data retention period		

Appendix C

Initial Utility Computing Reference Model Assessment

In this book, the Utility Computing Reference Model has been described in some detail. A key step in transforming an IT environment towards the characteristics of this model is the assessment (or benchmarking) of a current IT environment against the reference model standards. Only once this has been accomplished can one start to plan the utility transformation itself. In this section, we have compiled a questionnaire that should help a project team to assess a current IT estate quickly against the Utility Computing Reference Model specifications and 'score' the estate against this benchmark. This will then allow the project team to discuss and plan priorities for transformation and also to set the ultimate goals for a transformational initiative.

The questionnaire shown in this section is not designed to provide detailed analysis of the current state of an IT estate, but rather to give a high-level understanding as to what a project team should expect to have to transform in order to become compliant with the utility model. Often, this quick 'fact-finding' technique can be used to help decide which IT estate should be targeted first for transformation.

Delivering Utility Computing.   Guy Bunker and Darren Thomson
© 2006 VERITAS Software Corporation. All rights reserved.

Note

You do not necessarily have to do this for all of IT, you can approach this in terms of various potential services, rather than the ITO as a whole. For example, how does backup measure up to being turned into a utility, or storage, or servers or even applications or higher-level services such as disaster recovery?

Normally, this quick 'first pass' assessment would be followed by a more in-depth analysis, which would look at the component segments of each layer of the Utility Computing Reference Model and assess the targeted estate against each of these, in detail. This effort would require further time and effort in order for an appropriate scorecard and questionnaire to be developed, and for internal IT staff to be scheduled to provide the data necessary to complete this type of exercise.

The authors would be happy to discuss their experiences in developing detailed assessment scorecards with readers of this book, and can be contacted, via the publisher, at http://www.wiley.com.

The questionnaire included in this section uses a simple scoring system that can be extrapolated into a spreadsheet, or similar tool, in order for the user to gain insight into where inefficiencies or deficiencies exist within a current IT environment. Essentially, if the IT estate that is being assessed scores a '5' in every category within a given reference model layer, then it is likely that that it is already running as a utility and is compliant with the Utility Computing Reference Model. Each layer of the reference model has a series of associated questions that should be answered when completing this exercise. To score each layer of the reference model, a simple process of generating an average from the scores can be adopted.

Our suggestion is to set up meetings with relevant IT staff that are responsible for the targeted IT infrastructure and to work in partnership with these individuals in order to agree the scores associated with the targeted estate. In terms of using and presenting the data from this kind of assessment, we have found it useful in our projects to use standard spreadsheet graphical formats (such as Kiviat or 'Radar' charts) to represent the outcome of this type of assessment within management reports and business plans (Figure C.1). A blank version of this diagram, on which you can plot your company utility assessment, is given in Figure C.2 at the end of this appendix.

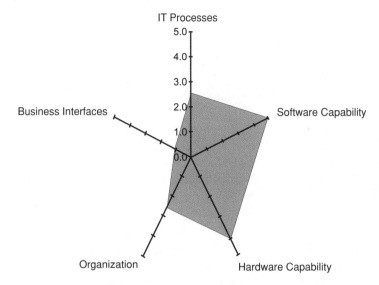

Figure C.1   Example data output from utility assessment.

C.1   UTILITY REFERENCE MODEL ASSESSMENT

C.1.1   The service layer

Context

Adopting a utility model for IT requires IT facilities to be delivered as services. Making users aware of the portfolio of IT services is achieved through a services catalog. In addition, consumers and suppliers of the IT utility service should be managed proactively so that maximum levels of consumer satisfaction and supplier alignment can be attained.

Question–Which of the following statements most accurately describes the current status of your IT services catalog?	**Score**
IT facilities are not yet delivered as services.	1
We have a list (document) that is available from the helpdesk that describes the main services offered by IT.	2
We have an online catalog where users can lookup what IT services are provided.	3

We have an online catalog where users can lookup and request an IT service.	4
We have a complete self-service portal where users can lookup and request an IT service, check their usage, confirm entitlement, etc.	5

Question–Which of the following statements most accurately describes the current method of consumer/supplier management?	**Score**
Our IT consumers and suppliers are managed on an adhoc, reactive basis. No formalized interface method exists for the communication and interaction with these parties.	1
Our IT consumers and suppliers are managed on an adhoc, reactive basis. However, a formalized interface method exists for the communication and interaction with these parties.	2
We are still managing suppliers and consumers with a traditional IT model but have built and use some of the management disciplines that exist within the Utility Computing Reference Model.	3
We treat IT users as consumers and are proactive in our supplier management technique but communication with these parties is reactive today.	4
Our consumers and suppliers are managed proactively and professional disciplines exist to manage all sales, marketing and billing activities. Service delivery standards are measured regularly and reported to all relevant stakeholders.	5

Question–Which of the following statements most accurately describes the current method of service portfolio management?	**Score**
No IT service portfolio exists. All IT services are currently provided on a reactive basis and we do not maintain a detailed service database of any kind to keep track of our service offerings or their popularity.	1
We keep a record of the IT services offered using spreadsheets (or similar tools). This record is	2

modified when new services become available or are retired, but no provision is made for tracking service usage. No proactive analysis occurs to gain input from consumers as to the relevance of the current offerings.	
All IT services are managed as part of a service portfolio. This portfolio is always kept current and its detail is available to all authorized users via a centralized portal. No proactive analysis is undertaken to ensure that the service portfolio is relevant and appropriate to consumer needs. No class of service model exists to give consumers choices in terms of service level characteristics.	3
All IT services are managed as part of a service portfolio. This portfolio is always kept current and its detail is available to all authorized users via a centralized portal. No proactive analysis is undertaken to ensure that the service portfolio is relevant and appropriate to consumer needs. The services that exist are offered with differing classes with prices that are appropriate to each class of service.	4
All IT services are managed as part of a service portfolio. This portfolio is always kept current and its detail is available to all authorized users via a centralized portal. Proactive analysis is undertaken to ensure that the service portfolio is relevant and appropriate to consumer needs. The services that exist are offered with differing classes with prices that are appropriate to each class of service.	5

C.1.2   The process layer

Context

A utility model for IT is predicated on having clearly defined 'take action' processes for enacting IT services. Without well-defined and appropriately automated processes, an IT utility will not run efficiently and service quality may be compromised.

Question–Which of the following statements most accurately describes the current status of your IT processes overall?	Score
We have an operational procedures document that describes some of our key processes.	1
All operational procedures are documented fully and this document is kept up to date and followed at all times.	2
We have started to adopt ITIL (or similar standard) for our key processes. We have documented workflows for offering and operating our IT services that are cross-functional.	3
We are fully compliant with ITIL (or similar standard) and we manage some automated workflows actively through a workflow engine.	4
Our workflows are automated to the maximum extent practical and we monitor status and exceptions actively. We are fully compliant with ITIL (or similar standard).	5

Question Which of the following statements most accurately describes the current maturity of your IT process lifecycle?	Score
We do not currently structure our IT management processes or think of them as a 'lifecycle'. Our defined processes are focused purely on the day-to-day running of IT and are invoked on a reactive basis.	1
We do not currently structure our IT management processes or think of them as a 'lifecycle'. Our defined processes are, however, documented fully and we have started to consider how to standardize processes to cover the lifecycle of our existing IT services.	2
We have now standardized and documented all of our IT processes that span the life of an IT service, from its planning to the testing of the final offerings. Our standards are compliant with ITIL (or similar standard).	3

We have now standardized and documented all of our IT processes that span the life of an IT service, from its planning, through its iterative assurance, ongoing management reporting and service redefinition/retirement. Our standards are compliant with ITIL (or similar standard) but we have not started to automate our processes at this stage.	4
We have now standardized and documented all of our IT processes that span the life of an IT service, from its planning, through its iterative assurance, ongoing management reporting and service redefinition/retirement. Our standards are compliant with ITIL (or similar standard). We have introduced a workflow engine to manage the automation of processes where this is appropriate.	5
Question—Which of the following statements most accurately describes the current mapping between your IT service lifecycle and your IT organizational structure?	**Score**
Our IT processes are not currently mapped to our IT organizational structure. Processes are executed by members of staff who have the basic capabilities required and who have the time to carry out a process at any given time.	1
We have a fairly well-defined organizational structure and, for the most part, IT processes are executed and managed by a relevant part of the IT organization. We often experience issues in the 'handover' as one part of the organization finishes its process and another is due to start the next.	2
Having defined and documented our IT process workflows, we are now in the process of mapping these to our organization and are undergoing the restructure of our IT group in accordance with this effort.	3

We try to ensure that every part of our process workflow is owned by a relevant center of excellence. However, we sometimes find that (perhaps due to staff shortages) some parts of our process lifecycle are not owned and managed as well as they could be.	4
Every part of our iterative IT process lifecycle is clearly mapped to a center of excellence within the organization. This is documented clearly, as are all of the interface points between the various parts of the organization. One part of our workflow links seamlessly with the next and we rarely experience issues caused by lack of ownership or responsibility.	5

C.1.3   The organizational layer

Context

To neglect the appropriate restructure of the IT organization during utility computing transformation would be to put the entire initiative at risk. An appropriate IT organization should have all of the relevant competencies and disciplines that are required to run an IT utility. The culture of the IT organization should match the 'consumer-oriented' principles that the utility model represents.

Question– Which of the following statements most accurately describes the current mapping between your IT organization and the centers of excellence described in the Utility Computing Reference Model?	Score
Our IT organization does not currently use a center of excellence model and the competencies that exist are grouped rigidly into the departments that you would expect to see within a traditional IT group (i.e. engineering, support, operations and strategy).	1

Our IT organization does not currently use a center of excellence model, but the competencies that exist to manage IT in a traditional way work well as collective teams and issues of ownership are rarely experienced.	2
We have started the move towards a center of excellence model, but do not believe that our current competencies cover any more than 40% of those defined within the UC Reference Model.	3
Our IT organization now contains centers of excellence that map to most of the disciplines shown in the UC Reference Model. We are reviewing the use of 'virtual', 'departmental' and 'outsourced' methods to drive further efficiencies into the organization.	4
Our IT organization now contains centers of excellence that map to all of the disciplines shown in the UC Reference Model. We use a combination of 'virtual', 'departmental' and 'outsourced' methods to accomplish this and are confident that this structure is as efficient as it can be.	5

Question–Which of the following statements most accurately describes the current culture of your IT organization?	**Score**
The culture of our current IT organization is one of 'technology guardian'. Our staff do not think of IT as a service, and they feel that their jobs are done well as long as 'the systems are up and running'.	1
Some elements of our organization have started to think about IT users as 'consumers'. This is driven largely by our strategy team, who are attempting to drive IT towards the utility approach. However, our IT personnel generally are cynical towards this strategy.	2
We now have a well-understood utility computing strategy, which our IT organization seems to be very positive about. All levels of the organization seem to understand the importance of a 'consumer-centric' culture, although no actual cultural change has been experienced yet.	3

We now have a well-understood utility computing strategy, which our IT organization is very positive about and is keen to drive forward. All levels of the organization understand the importance of a 'consumer-centric' culture and we are already seeing the benefits of this through closer relationships and understanding with IT consumers.	4
Our entire IT organization is now very 'consumer focused' and it considers itself to be a 'service provider'. All members of staff consider the consumer before all others, and a generally positive and excited atmosphere now exists within IT. We are proud of the services that we provide and maintain excellent consumer relationships throughout.	5

Question–Which of the following statements most accurately describes the current relationship between your IT organization and its service consumers?	**Score**
We experience constantly a mismatch between our IT users needs/expectations and what we are able to provide them. This has led to a poor relationship between IT and its consumers.	1
Our relationship with our IT consumers is fairly positive but we always seem to be playing 'catch up' and never seem to have enough money or resources to respond to business demand.	2
We have a better relationship with some of our consumers than others. A much more 'consumer-centric' model has been introduced to deliver some discrete areas of IT service, and in these areas our relationship with consumers is good.	3
We have now adopted generally a 'consumer-centric' approach to delivering IT services and this means that we understand our consumers' demands before time and are able to respond proactively to business change. Our relationship with our consumers is now changing and becoming more positive than ever before.	4

Our 'consumer-centric' approach to delivering IT services now means that we understand our consumers' demands before time and are able to respond proactively to business change. As a result, our relationship with our service consumers is extremely positive and is getting better every day. Both the IT department and its consumers feel that they are operating in total alignment.	5

C.1.4   The software layer

Context

Innovative use of software is key in realizing a utility computing vision. A successful utility computing implementation will rely heavily on the ability to abstract the complexities of hardware configuration and dynamic asset provisioning, and the detailed and efficient reporting of all facets of the IT infrastructure and associated service levels.

Question–Which of the following statements most accurately describes the reporting capabilities that exist within the targeted IT estate?	**Score**
The current system provides no facility for the reporting of process or function to the system user.	1
Some basic reports can be created from the current system. Reports can be produced by analyzing system logs manually and by documenting these in an appropriate format.	2
Some basic reports can be created from the current system. Reports are produced by bespoke scripts that analyze the system logs automatically and produce output in an appropriate format.	3
The current system provides reports detailing most processes and functions, these reports are not configurable but are generated automatically and delivered to appropriate users of the system.	4

The current system provides reports detailing all appropriate processes and functions, these reports are fully configurable and are generated automatically and delivered to appropriate users of the system.	5

Question–Which of the following statements most accurately describes the method of system access and security adopted by the current IT management system?	**Score**
The current system does not allow IT elements to be accessed remotely. The system does not comply with a recognized security standard.	1
The current system supports remote system access to some named servers. These connections do not comply with a recognized security standard.	2
The current system supports remote system access. All servers can be accessed in this way. Connections can be made by authorized users using industry standard web browsers.	3
The current system supports remote system access. All elements of the server and networking estates can be accessed in this way. All connections are secure (and comply with a recognized security standard) and can be made by authorized users using industry standard web browsers.	4
The current system supports remote system access fully. All elements of the server, storage and networking estates can be accessed in this way. All connections are secure (and comply with a recognized security standard) and can be made by authorized users using industry standard web browsers.	5

Question–Which of the following statements most accurately describes the automated discovery and inventory capabilities of the current IT management system?	**Score**
The current system is unable to create inventories of networked IT assets of any kind.	1
The current system provides some basic inventory management functionality and is capable of automatically creating and managing inventories containing most major elements of networking, storage and server hardware.	2

The current system provides some basic inventory management functionality and is capable of creating and managing inventories containing most major elements of networking, storage and server hardware. In addition, common applications can also be captured.	3
The current system provides full inventory management functionality and is capable of automatically creating and managing inventories containing all major elements of networking, storage and server hardware. In addition, all common O/Ss and applications can also be captured.	4
The current system provides full inventory management functionality and is capable of automatically creating and managing inventories containing all major elements of networking, storage and server hardware. In addition, all O/Ss, applications, associated software elements (such as patches) and 'bare metal' systems can also be captured.	5
Question–Which of the following statements most accurately describes the automated asset provisioning and life cycle management capabilities of the current IT management system?	**Score**
The current system is unable automatically to manage networked IT assets of any kind.	1
The current system provides some basic automotive management functionality and is capable of basic software installations, configuration updates and upgrades for servers.	2
The current system provides some basic automotive management functionality and is capable of basic software installations, configuration updates and upgrades for servers, storage and networking components.	3
The current system provides fully automated management functionality and is capable of all software installations, configuration updates and upgrades for servers, storage and networking components.	4

The current system provides fully automated management functionality and is capable of all software installations, configuration updates and upgrades for servers, storage and networking components. 'bare metal' system restores are also possible currently.	5

Question–Which of the following statements most accurately describes the hardware virtualisation capabilities of the current IT management system? — **Score**

The current system is unable to virtualize any of our hardware assets. As a result, the human effort involvement in managing these assets is unacceptable and provisioning times are too high, which impacts on our business. — 1

The current system provides some basic virtualization functionality (perhaps in the management of virtual storage devices). As a result, we have started to see some reduction in IT management cost. — 2

The current system provides virtualization across most elements of the targeted hardware estate. We are able, as a result, to create hardware asset 'pools' in some discrete areas. The costs associated with IT management (particularly for the parts of the estate that remain non-virtualized) remain unacceptable at this stage. — 3

The current system provides virtualization across all elements of the targeted hardware estate. However, we have not yet adopted automated provisioning techniques and still rely heavily on human intervention. As a result, the costs associated with IT management remain unacceptable at this stage. — 4

The current system provides virtualization across all elements of the targeted hardware estate. We are able, as a result, to create hardware asset 'pools' and provision hardware automatically on a 'just in time' basis. The costs associated with IT management are acceptable to us. — 5

Question–Which of the following statements most accurately describes the 'root cause analysis' capabilities of the current IT management system?	Score
The current system provides no facility for conducting root cause analysis. As a result, we find it difficult to resolve issues swiftly in system performance or availability.	1
Some basic root cause analysis can be undertaken within the current system. This can be executed by analyzing current data and by process of 'trial and error'.	2
Some basic root cause analysis reports can be created from the current system. Reports are produced by bespoke scripts that analyze current data automatically and produce output in an appropriate format.	3
The current system provides the facility to create root cause analysis reports. These reports are not configurable, but are generated automatically and delivered to appropriate administrators of the utility environment. Most problems are resolved within acceptable timeframes as a result.	4
The current system provides full root cause analysis reporting. Reports are fully configurable and are generated automatically and delivered to appropriate system administration staff. All problems are resolved within acceptable timeframes as a result.	5

C.1.5   The hardware layer

Context

A key goal of a utility computing strategy is to drive the utilization rates associated with IT hardware asset up and, wherever possible, to reduce the cost of hardware procurement through commodization and consolidation. At the same time, hardware assets represent the source of raw power and capacity on which the utility will rely. To this end, it is vital that the hardware used to 'fuel' an IT service is designed and configured appropriately.

Question–Which of the following statements most accurately describes the current hardware utilization levels that exist within the targeted IT estate?	Score
Utilization of the major IT assets that form this system is at a unacceptable level (typically below 15%).	1
Utilization of the major IT assets that form this system is generally low (typically between 15 and 40%). Major improvement to ROI could be made.	2
Utilization of the major IT assets that form this system is at an acceptable level (typically around 40–60%). Peaks in demand have been known to cause performance issues.	3
Utilization of the major IT assets that form this system is at an acceptable level (typically around 40–60%). Peaks in demand do not cause performance issues.	4
Utilization of the major IT assets that form this system is always well tuned and utilized (typically around 60–80%). This demonstrates excellent ROI from our perspective and no risk of asset constraint.	5
Question–Which of the following statements most accurately describes the current scalability characteristics of the targeted IT estate?	Score
System scalability is a cause for major concern in this area. Constant system capacity/performance issues are experienced as system load is increased.	1
System scalability has caused some issues to the business. In addition, the IT organization is not confident that, with increased system load, the current system will scale appropriately and without disruption to business.	2

System scalability is not currently an issue in this area. However, the IT organization is not confident that, with increased system load, the current system will scale appropriately and without disruption to business.	3
System scalability is not currently an issue in this area. However, the IT organization has only a reasonable level of confidence that, with increased system load, the current system will scale appropriately and without disruption to business.	4
System scalability is not an issue in this area and the IT organization has every confidence that, with increased system load, the current system will scale appropriately and without disruption to business.	5

Question–Which of the following statements most accurately describes the current resilience characteristics of the targeted IT estate?	**Score**
The system is suffering constantly from either planned or unplanned 'downtime'. The amount of system outage experienced is totally unacceptable.	1
The system is suffering frequently from either planned or unplanned 'downtime'. System availability in this area should be improved.	2
The system demonstrates a reasonable level of availability. There have been cases where system 'downtime' has caused business issues.	3
Issues are rarely experienced with regard to system availability in this area. The system is resilient to most types of infrastructural failure.	4
No issues are ever experienced with regard to system availability. The system is totally resilient to all types of infrastructural failure.	5

Question–Which of the following statements most accurately describes the current performance characteristics of the targeted IT estate?	Score
System performance is a cause for major concern in this area. Constant performance issues are experienced by users and this is affecting consumer satisfaction adversely.	1
System performance is a cause for concern in this area. Performance issues are experienced by users very often.	2
Users of the system generally experience acceptable performance, but issues are experienced on a regular basis.	3
System performance is rarely an issue in this area. Users of the system usually experience appropriate, or more than sufficient, system performance.	4
System performance is not an issue in this area. Users of the system always experience appropriate, or more than sufficient, system performance.	5

Question–Which of the following statements most accurately describes the centralized nature of the targeted IT estate?	Score
The infrastructure is very distributed. This causes major problems in managing and controlling the targeted IT estate.	1
The system is mostly distributed in its design. This causes some problems in managing and controlling it.	2
The system is mostly centralized in its design. This enables reasonably effective management and control.	3
Issues are rarely experienced with regard to system availability in this area. The system is resilient to most types of infrastructural failure.	4
The system is fully centralized in its design. This enables very effective management and control.	5

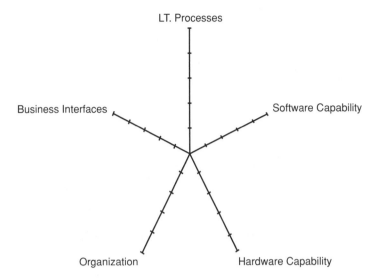

Figure C.2 Blank diagram on which to plot your utility assessment.

C.2 PLOTTING YOUR UTILITY ASSESSMENT RESULTS

Figure C.2 provides a blank version of the disgram shown earlier in Figure C.1. This diagram may be used to plot the results obtained from your utility assessment questionnaire.

Appendix D

Stakeholders and Objections

Identification of the key stakeholders in any utility computing project will be essential to its success. Below is a list of the potential stakeholders and some of their typical concerns with delivering IT as a service. Understanding the people and addressing their personal concerns and needs will help to build a team, which, in turn, will deliver a successful utility.

Note
When embarking on a utility computing initiative, there will be objections from many areas of the organization. Developing a Frequently Asked Questions document and publishing it will help reduce the fear, uncertainty and doubt that users may have. Where possible, try to avoid acronyms as they can be misunderstood or misinterpreted easily.

CEO	
Too much hype	Need to prove ROI and that the new method of doing things is a worthwhile investment.

Delivering Utility Computing. Guy Bunker and Darren Thomson
© 2006 VERITAS Software Corporation. All rights reserved.

CIO

Too much hype	Similar to the CEO, need to prove the ROI.
	Proof of other efficiencies is also important, for example, customer satisfaction surveys.
	Proof of utilization as a means to justify additional investment in resources.
More management software? I've already bought more than I can use	Are there things that might be reduced? For example, consolidating virtualization software across all platforms?
Too expensive	Can you prove that the introduction of new software will result in deferred hardware purchases? For example, increasing server utilization through colocation of applications or use of virtual machines.

Administrator

All this process is slowing me down	Need to prove efficiency. Process workflows are a simple way to monitor responsiveness.
There has been no reduction in headcount	Are you able to do more with less? How much more storage, or more servers do you manage?
	Can you respond to business requests more efficiently?
There has been a reduction in headcount, you promised this would not happen	Is it really a reduction, or just a change in required skills?
	Where have the reductions been? Is it in operations staff, or in administrators?

	Are you able to manage more effectively with the new processes?
	Would you have reduced the staffing level?
Fewer people are using this service than when we started it	Has the service outlived its usefulness or been replaced by a different one? Can you remove it entirely?
Move things around automatically? That will just make things worse, it is complicated enough as it is	Monitoring tools are essential – ideally these are both graphical and report-based.
	Automation tools can give the option to move applications fully automatically or semi-automatically, so the administrator need not feel that they are losing control.
There are no two systems alike, and I have 1000s of them, what chance is there of standardization?	This is an attitude of mind and will require support from the top. By standardizing solutions, it will be both easier and cheaper to support them. The goal is not to get to a single service solution, but rather a handful of services.
Template solutions, how boring – I like a challenge	The good news is that with a template solution there can be improved testing and fewer reactive calls. The result is more time for research into new technology that can be put into innovative new services.
Innovation . . . I do not have enough time for that	Standardization actually creates time. There are fewer oddities that need to be resolved and so there is time to do more interesting things, such as looking at new technologies.

We already have a process for configuring new servers/storage	Good. Existing processes can be used in a utility computing initiative and might be able to be augmented with additional technology, such as process workflow, to make it even simpler and more consistent.
Another efficiency program? I'm already up to my neck in consolidation programs	The principles behind utility computing are a good match for a consolidation initiative. It would be worth looking at the various principles and processes to see if they could be used to assist in the consolidation program.
Billing? People will not like it	Money is always an emotive subject. Often, businesses do not want to charge the lines of business for their IT usage, more likely, they just want to give visibility (sometimes known as showback) into the costs involved. This will help them determine whether the IT investment made is a good business decision.
Understand the business? That is not my job	In the new ITO, it is essential for everyone to understand, at some level, the business. If the ITO is to be a customer-driven service organization, then understanding the needs of the customer is of paramount importance.
Operator With all this automation, will I lose my job?	No. Far from it, process automation will enable you to do more for the business and carry out tasks that previously were done by the administrator.

I am concerned about messing things up with this workflow stuff

Workflow is designed to remove the risk from operations such as provisioning new servers or storage. It enables you to follow a process, automating much of it where possible, and gives 'options' where required.

What happens if they request something that I cannot deliver?

In theory this should not happen! Reporting will enable the ITO to understand its current position with respect to allocated and unallocated resources. There is then the ability to match resources to services to determine what is required should a service be requested. Obviously, if there is no resource to allocate, then the service as requested cannot be fulfilled. However, it may be that a reduced service could be provisioned and the complete one created later when new resources are found. Discussion with the user as to what is, and what is not, acceptable is of paramount importance in this situation – and the sooner the discussion, the better!

Line of business: application owner

I want to keep my servers, I paid for them

Asset ownership is one of the biggest barriers to utility computing! Success will require a change in thinking from 'what hardware is my application running on?' to 'how is my application running?' By specifying how the application should run: response time, number of users, etc., and leaving the hardware decision to the ITO, a more cost-effective and efficient business solution can be created.

I do not want to use any of the template solutions provided – they do not fit my requirements	Why not? The ITO is a customer-driven organization, it should be able to help you select which solution is right for you. If there really is none that fits, and there are others with similar requirements, then you could help them design a new service that does fit your needs. However, the ITO will not create custom solutions for each and every application, as this will defeat some of the utility principles that it is using.

Line of business: application developer

More process? We are supposed to respond quickly!	This is not about more process – although there will be some new process. In theory it will be easier, like at a restaurant, it will be possible to choose a configuration from a menu and have it delivered for use more rapidly than at present.
Usage statistics? I have not got time to add that into my development plans	By thinking about usage when developing an application, it will be easier to understand how performant the IT environment is. It might just be a simple log for transactions, or maybe information on throughput – all could be used to help determine a service level agreement.

Line of business: user

The new system is worse than before	The need for service level agreements and accountability, especially for performance, becomes apparent. It is essential to be able

to prove that the new environment is no worse than before – and is, hopefully, better.

This is also true when upgrades happen, whether it is from a new patch release of the OS, or totally new hardware.

Support

This self-service way of working is causing me more problems than before

Why? Often a new way of working will cause problems because it is different, rather than because it is worse. Are there things that can be done to alleviate the issues, such as some simple training or education into why the new procedures are in place? New, or adapted, process may also be required to ease the transition.

Suppliers

It is all very well asking for on-demand pricing, but we do not support this concept

The first question is why not? There may be alternatives that are supported, or other arrangements that can be made, such as having rapid response to requests.

If the supplier understands the configurations that the ITO is supporting and the SLAs that are proposed, then maybe they can consider holding stock specifically for the customer and delivering it in two days, rather than six weeks.

Something similar for software licensing could also be arranged, or perhaps the introduction of term-based licensing, where the software is only to be used for three or six months rather than a number of years.

Upfront discussion with the supplier will be needed to create the utility computing environment.

Glossary

As with any industry, and especially with new paradigms, there is terminology. Often, different interpretations of common (or overused) terms can cause confusion. By starting out with a common understanding of terms, many problems and issues can be avoided.

Note
When embarking on a utility computing initiative, it is worth creating a glossary for the project. If a term arises that causes confusion, write down and agree upon a definition. Beg, borrow or steal other terms and definitions and post them on a website, where they can be found easily by all involved.

Administrator *See* IT Administrator

Autonomic computing An IBM initiative to create a self-managing IT infrastructure.

Availability The amount of time that a system or application is available during those time periods when it is expected to be available. Often measured as a percentage of an elapsed year. For example, 99.95% availability equates to 4.38 hours of downtime in a year ($0.0005 \times 365 \times 24 = 4.38$) for a system that is expected to be available all the time.

Blade computing *See* Blade server

Delivering Utility Computing.   Guy Bunker and Darren Thomson

Blade server   A small form factor computer system with one or more processors, memory, storage and network. Often known as 'blades,' these can be grouped into racks of tens or hundreds of servers to provide high performance in a compact space. Often used when servers are required to be provisioned and reprovisioned repeatedly.

BPEL   *See* Business process execution language

BU   *See* Business unit

Business process execution language   A standard to define business process behavior, for example, sequencing of activities or the correlation of messages and processes. Based on web services.

Business unit   Synonymous with Line of business.

Chargeback   The process of billing an internal user for use of resources based on usage. While not often carried out, the bill is created to give the user visibility into their usage and, therefore, associated costs. Sometimes referred to as showback.

CIO   Chief Information Officer. Usually a board-level executive responsible for all IT and IT infrastructure within an organization. Also known as the IT Director.

Class of service   The set of service level objectives with their metrics. Used to implement one or more service level agreements. Grouping mechanism.

DBA   Database Administrator. Responsible for all aspects of a database application, for example, installation, configuration and management.

Ethernet   The predominant local area network (LAN) technology.

Host Bus Adaptor   An i/o adaptor to enable the transfer of information between the host computer bus and fibre channel storage. Often used in SAN's.

Information Technology (IT)   A collective term for equipment, software and other facilities used to process electronic information or provide information services.

IT Administrator   A person responsible for installation, configuration and management of a computer system and attached peripherals. Often responsible for business applications as well.

IT Director *See* CIO

IT Operator Responsible for the day-to-day running and management of computer systems and attached peripherals. Spends time watching alert consoles and reacting to events rather than installation and configuration tasks. Synonymous with IT Administrator in smaller organizations.

IT organization Responsible for the operation of production information systems (and often development systems) within an organization. Responsible for overall IT architecture design and implementation to ensure lines of business IT needs are met. Within utility computing, the role is changing to become more service-oriented and requires new skills for customer relationship management.

ITIL IT Infrastructure Library. Developed for the British government, it is now the defacto standard across the world. Delivered as a series of best practice guidelines for service management from planning to provisioning and support of IT services.

ITIL Information Technology Infrastructure Library. A set of best practice standards for service management within information Technology (IT). Originally developed for the British Government it is now a de-facto global standard. www.itil.org

ITO *See* IT organization

Line(s) of business One or more divisions within an organization, responsible for delivering business value. Use(s) the ITO to run the information systems required for their business.

LOB *See* Line of business

Logical unit (number) A partial address by which a virtual storage device is recognized and accessed on an i/o bus or storage network.

LUN *See* Logical unit (number)

OLA *See* Operational Level Agreement

Operational Level Agreement Similar to Service Level Agreement but, typically, held between two technical functions (as opposed to a technical and business function)

Operator *See* IT operator

PAW *See* Process automation workflow

Policy   Affects the behavior of the system. Can apply to one or more classes of object with something in common, as well as individual objects. User configurable. Programmable at a different time from compile time, as opposed to 'algorithm'. Can be grouped into hierarchies. Generic term for policy rules and policy groups. Not all configuration is policy, but policy can be configuration.

Synonym for 'Policy definition':

- *Internal policy* controls self-management of the software product.
- *External policy* controls behavior of managed resources.
- *Low-level policy* – more procedural and concrete than a high-level policy. Specifically targeted. Usually expressed as one or more policy rules. One or more condition + action rules. *See also*: Policy action, Policy condition, Policy rule. Synonym for Executable policy.
- *High-level policy* – policies that are more declarative in nature. Expressed in terms of goals rather than actions. Can have broader targets, more abstract goals/objectives. Can be implemented or realized through lower-level policies. A policy expressed in terms of service level objectives is an example of one kind of high-level policy.

Policy action   Definition of what needs to be done to enforce a policy rule when the conditions of the rule are met. Can be a method invocation, a system command, a script call or a notification/ alert.

Policy-based automation   Automation driven by predefined policies. A technology that enables scalable and lights-out management of IT environments via the definition of policies.

Policy cache   Synonym for Policy repository. *See also*: Policy source.

Policy condition   Representation of the necessary state(s) and/or prerequisite(s) that define whether a policy rule's actions should be performed.

Policy decision point   Logical entity that makes policy decisions.

Policy domain   Collection of managed resources and services over which a common and consistent set of policies are administered in a coordinated fashion. Synonym for Policy scope.

Policy enforcement point   Enforces policy decisions made by the decision point.

Policy group   A mechanism for hierarchical organization of policies. A container aggregating policy rules or other policy groups. Contains one or more policies. Not all policy groups are high-level policies. *See also* Policy.

Policy repository   A logical container for all the policies in a particular domain. This allows access to policies, without having to know exactly which policy source contains them. Some implementations may also allow for caching of some of the information about a policy (e.g. most frequently used information), rather than just providing a reference to the policy in the particular policy source. Note: a particular host may support the policy source and policy repository interfaces, or just one of them, for a particular policy domain. Synonym for Policy cache.

Policy rule   Basic building block of the policy system. One or more conditions and one or more actions to perform if the conditions evaluate 'true'. Synonym for Rule.

Policy rule engine   The system component that evaluates the policies. The system may have multiple, specialized and/or distributed policy rule engines.

Policy source   Primary store of policy data for a particular policy domain. This is the machine with the attached physical storage that contains the policy data. All updates and changes are processed here, and feed from here to replicas or caches. *See also*: Policy cache, Policy repository.

Procedural policy   Synonym for Low-level policy (procedural/ executable policy).

Process automation workflow   The description of a series of process steps which may carry out a task fully or semi-automatically. Frequently used in provisioning of IT services and policy actions.

Process step   An individual component step that can be assembled into a workflow.

Process workflow engine   The system component that evaluates and executes process automation workflows. It may have multiple distributed agents for executing process steps on remote resources.

Rate card   Simple description of services (usually limited to a single area, for example servers), lists characteristics of the service and cost associated with each. *See also* Service catalog.

Returns on Investment   The monetory benefit gained by spending money to develop or improve a system or process.

SAN   *See* Storage area network

Sandbox   A protected IT environment where new ideas can be tried without risk to the rest of the environment.

Service   In the context of IT, some well-defined functionality provided to one party by another, and governed by the terms of a service level agreement. The consumer and provider are generally in different organizations, and may be in entirely different companies.

Service catalog   A list of services available from the ITO to the customer. Often implemented through a web interface and displaying cost (chargeback) information to help the customer decide which service is most applicable to them.

Service level   Quantification of quality of service as measured through specific predefined metrics.

Service level agreement   The contract between the provider and consumer of a service which governs delivery of the service. Specifies attributes of the service, such as required levels of availability, serviceability, recoverability and performance. Requirements are often expressed in business terms, and are referred to as service level objectives. Penalties can be imposed for breaches of the SLA.

Service level expectation   Often used in place of a service level agreement, especially if there are no penalties for non-compliance.

Service level management   Process of managing IT as a service to a particular Quality of Service (QoS). Often implies automation for provisioning a service to a specific level and monitoring it to ensure that it remains within the required boundaries. Automation can also be used to correct service level breaches, for example adjusting the configuration or increasing resources available for a service. Often used in conjunction with service level agreements (SLAs).

Service level objective   An individual service goal expressed in a measurable way. Defines a metric and a threshold, or range, for the metric, used to enforce and/or monitor SLA compliance.

Generally declarative. The objective is met if the metric (as measured or calculated) value is within the acceptable range.

Service-oriented architecture   A collection of services which communicate with each other. Services can be internal or external to the enterprise utilizing the service-oriented architecture.

Showback   *See* Chargeback

Simple object access protocol   A lightweight XML-based protocol to exchange information in a distributed environment. Often used in conjunction with web services.

SLA   *See* service level agreement

SLE   *See* service level expectation

SLM   *See* service level management

SLO   *See* service level objective

Short Message Service A service for sending short messages between cellular phones.

SOA   *See* service oriented architecture

SOAP   *See* simple object access protocol

Storage area network   A network whose primary purpose is the transfer of data between computer systems and storage devices, or between storage devices.

Total Cost of Ownership   A calculation to help assess both direct and indirect costs associated with an item.

Utility   A business that provides standardized essential services reliably enough that they are taken for granted by customers. Services are metered and the customer is billed based on usage.

Utility computing   The technology, tools and processes that collectively deliver, monitor and manage IT as a service to users.

Utility management   The final stage in the transition from a traditional to a utility computing model. Automation enables dynamic deployment of resources to optimize utilization based on needs, and from a business alignment standpoint, the transition to a self-service user mentality should be complete. Accountability, including chargeback, should be mature.

Virtualization   Abstraction of key properties of a physical device and presentation of those properties to users of the device. In IT, this includes servers, network and disk and tape storage. It is most commonly implemented using a software layer between the physical device, or devices, and the application that wants to use it.

Virtual local area network   Technology that enables a group of computer resources to behave as though they are connected to a single network segment, even if they are not.

VLAN   *See* Virtual local area network

Web services   A set of standards promoted by the W3C for working with other applications and organizations. Often includes SOAP and WSDL.

Web service description language   An XML format for describing network services. It operates on messages containing either document-oriented or procedure-oriented information.

Workflow   *See* Process automation workflow

WSDL   *See* Web service description language

XML   Extensible markup language. A metalanguage used to allow easy exchange of information. Especially across the web.

Index